Undergraduate Texts in Mathematics

Undergraduate Texts in Mathematics

Undergraduate Texts in Mathematics are generally aimed at third- and fourth-year undergraduate mathematics students at North American universities. These texts strive to provide students and teachers with new perspectives and novel approaches. The books include motivation that guides the reader to an appreciation of interrelations among different aspects of the subject. They feature examples that illustrate key concepts as well as exercises that strengthen understanding.

For further volumes:
http://www.springer.com/series/666

John Stillwell

The Real Numbers

An Introduction to Set Theory and Analysis

 Springer

John Stillwell
Department of Mathematics
University of San Francisco
San Francisco, CA, USA

ISSN 0172-6056
ISBN 978-3-319-34726-4   ISBN 978-3-319-01577-4 (eBook)
DOI 10.1007/978-3-319-01577-4
Springer Cham Heidelberg New York Dordrecht London

Mathematics Subject Classification: 03-01, 26-01

To Elaine

Preface

Every mathematician uses the real number system, but mathematics students are seldom told what it is. The typical undergraduate real analysis course, which is supposed to explain the foundations of calculus, usually assumes a definition of $\mathbb{R}$, or else relegates it to an appendix. By failing to reach the *real foundation* (pun intended), real analysis runs the risk of looking like a mere rerun of calculus, but with more tedious proofs. A serious look at the real numbers, on the other hand, opens the eyes of students to a new world—a world of infinite sets, where the need for new ideas and new methods of proof is obvious. Not only are theorems about the real numbers interesting in themselves, they fit into the fundamental concepts of real analysis—limits, continuity, and measure—like a hand in a glove.

However, any book that revisits the foundations of analysis has to reckon with the formidable precedent of Edmund Landau's *Grundlagen der Analysis* (Foundations of Analysis) of 1930. Indeed, the influence of Landau's book is probably the reason that so few books since 1930 have even attempted to include the construction of the real numbers in an introduction to analysis. On the one hand, Landau's account is virtually the last word in rigor. The only way to be more rigorous would be to rewrite Landau's proofs in computer-checkable form—which has in fact been done recently. On the other hand, Landau's book is almost pathologically reader-unfriendly. In his *Preface for the Student* he says "Please forget everything you have learned in school; for you haven't learned it," and in his *Preface for the Teacher* "My book is written, as befits such easy material, in merciless telegram style." While memories of Landau still linger, so too does fear of the real numbers.

In my opinion, the problem with Landau's book is not so much the rigor (though it is excessive), but the lack of background, history, examples, and explanatory remarks. Also, the fact that he does *nothing* with the real numbers except construct them. In short, it could be an entirely different story if it were explained that the real numbers are interesting! This is what I have tried to do in the present book.

In fact the real numbers perfectly exemplify the saying of Carl Ludwig Siegel that the mathematical universe is inhabited not only by important species but also by interesting individuals. There are interesting individual numbers (such as $\sqrt{2}$, e, and π), interesting sets of real numbers (such as the Cantor set, Vitali's nonmeasurable set), and even interesting sets of which no interesting member is known (such as the set of normal numbers). All of these examples were known in 1930, but in recent decades they have been joined by many new exotic sets arising from the study of fractals, chaos, and dynamical systems.

The exotic sets arising from dynamical systems are one reason, I believe, to shift the emphasis of analysis somewhat from functions to sets. Of course, we are still interested in sequences of numbers and sequences of functions, and their limits. But now it seems equally reasonable to study sequences of sets, since many interesting sets, such as the Cantor set, arise as their limits. Another reason is simply the great advances made by set theory itself in recent decades, many of them motivated by the desire to better understand the real numbers. These advances are too technical for us to discuss in detail, but they result from the fundamental fact that analysis is based on uncountable sets and the struggle to understand this fact.

The set of real numbers is the first, and still the most interesting, example of an uncountable set. The second example is the set of countable ordinals. It is less familiar to most mathematicians, but also of great importance in analysis. If analysis is taken to be the study of limit processes, then *countable ordinals are the numbers that measure the complexity of functions and sets defined as limits of sequences*. In particular, we assign the lowest level of complexity (zero) to the continuous functions, the next level of complexity (one) to the functions that are not continuous but are limits of continuous functions, complexity level two to functions that are not of level one but are limits of functions of level one, and so on. It turns out that there are functions of all levels $0, 1, 2, 3, \ldots$ *and beyond*, because one can find a sequence of functions $f_0, f_1, f_2, \ldots$, respectively of levels $0, 1, 2, \ldots$, whose limit is *not* at any of these levels. This calls for a *transfinite* number, called ω, to label the first level beyond $0, 1, 2, \ldots$.

The transfinite numbers needed to label the levels of complexity obtainable by limit processes not only make up an uncountable set: in fact they make up the *smallest* uncountable set. Thus, the raw materials of analysis—real numbers and limits—lead us to two uncountable sets that are seemingly very different. Whether these two sets are actually related—specifically, whether there is a bijection between the two—is the fundamental problem about real numbers: the *continuum problem*. The continuum problem was number one on Hilbert's famous list of mathematical problems of 1900, and it still has not been solved. However, it has had enormous influence on the development of set theory and analysis.

The above train of thought explains, I hope, why the present book is about set theory *and* analysis. The two subjects are too closely related to be treated separately, even though the usual undergraduate curriculum tries to do so. The typical set theory course fails to explain how set concepts are relevant to analysis—even seemingly abstruse ones such as different axioms of choice and large cardinals. And the typical real analysis course fails to address the set issues that arise inevitably from the real numbers, and from measure theory in particular. When the two subjects are treated together one gets (almost) two courses for the price of one.

The book expands some of the material in my semi-popular book *Roads to Infinity* (Stillwell 2010) in textbook format, with more complete proofs, exercises to reinforce them, and strengthened connections with analysis. The historical remarks, in particular, explain how the concepts of real number and infinity developed to meet the needs of analysis from ancient times to the late twentieth century.

In writing the book, I had in mind an audience of senior undergraduates who have studied calculus and other basic mathematics. But I expect it will also be useful to graduate students and professional mathematicians who until now have been content to "assume" the real numbers. I would not go as far as Landau ("please forget everything you have learned in school; for you haven't learned it") but I believe it is enlightening, and fun, to learn something new about the real numbers.

My thanks go to José Ferreiros and anonymous reviewers at Springer for corrections and helpful comments, and to my wife Elaine for her usual tireless proofreading. I also thank the University of San Francisco and Monash University for their support while I was researching and writing the book.

San Francisco, CA, USA   John Stillwell

Acknowledgments

I would like to thank the following for permissions: the Dolph Briscoe Center for American History in Austin, Texas, for permission to use the photo of John von Neumann in Fig. 6.3; Don Albers and Jerry Alexanderson for permission to use the photo of Paul Cohen in Fig. 7.3; and Professor Alina Filipowicz-Banach for permission to use the photo of Stefan Banach in Fig. 9.2.

Contents

Chapter 1
The Fundamental Questions

PREVIEW

On the historical scale, analysis is a modern discipline with ancient roots. The machinery of analysis—the calculus—is a fusion of arithmetic with geometry that has been in existence for only a few hundred years, but the problem of achieving such a fusion is much older. The problem of combining arithmetic and geometry occurs in Euclid's *Elements*, around 300 BCE, and indeed Euclid includes several of the ideas that we use to solve this problem today.

In this preliminary chapter we introduce the basic problems arising from attempts to reconcile arithmetic with geometry, by discussing certain fundamental questions such as:

- What are numbers?
- What is the line?
- What is geometry?

The ancient Greeks discovered the basic difficulty in reconciling arithmetic with geometry, namely, the existence of *irrationals*. Irrationals are needed to fill *gaps* in the naive concept of number, and these gaps can only be filled by admitting infinite processes into mathematics. Thus, to develop a number concept complete enough for calculus, we need a theory of infinity. The development of such a theory will be the subject of later chapters.

1.1 A Specific Question: Why Does $ab = ba$?

This question is not as trivial as it looks. Even if we agree that a and b are numbers, and ab is the product of a and b, we still have to agree on the meaning of numbers and the meaning of product—and these turn out to be deep and fascinating issues. To see why, consider how ab was understood from the time of ancient Greece until about 1860.

J. Stillwell, *The Real Numbers: An Introduction to Set Theory and Analysis*,
Undergraduate Texts in Mathematics, DOI 10.1007/978-3-319-01577-4_1,
© Springer International Publishing Switzerland 2013

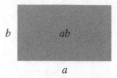

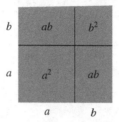

Fig. 1.1 The product *ab* of lengths *a* and *b*

Fig. 1.2 The rectangle picture of $(a + b)^2$

In Greek mathematics, and in Euclid's *Elements* in particular, quantities *a* and *b* were viewed as *lengths*, and their product *ab* was taken to be the rectangle with perpendicular sides *a* and *b* (Fig. 1.1.)

Then it is completely obvious that $ab = ba$, because the rectangle with perpendicular sides *b* and *a* is the same as the rectangle with perpendicular sides *a* and *b*. It was so obvious that Euclid did not bother to point it out and, to the Greeks, $ab = ba$ was probably not an interesting fact, because it was true virtually by definition.

This could be considered a virtue of the rectangle definition; it makes the basic algebraic properties of the product available at a glance, so that one does not need to think about them. One such property, which Euclid *did* point out, is the formula that we write as

$$(a + b)^2 = a^2 + 2ab + b^2.$$

Many a beginning algebra student thinks that $(a + b)^2 = a^2 + b^2$, but this mistake will not be made by anyone who looks at the rectangle picture of $(a + b)^2$ (Fig. 1.2). Clearly, the square with side $a + b$ consists of a square a^2 with side *a*, a square b^2 with side *b*, but also two rectangles *ab*. Hence $(a + b)^2 = a^2 + 2ab + b^2$. The Greeks were so fond of this picture that they even stamped it on coins! Figure 1.3 shows an example from the Greek island of Aegina, from around 400 BCE, even before the time of Euclid.

This is "algebra," but not as we know it. It runs alongside our algebra up to products of three lengths, but refuses to go further. The product of lengths *a*, *b*, and *c* was interpreted by the Greeks as a *box* with perpendicular sides *a*, *b*, and *c*. This interpretation agrees with ours—and makes it possible to visualize results such as

Fig. 1.3 Aegina coin

$(a + b)^3 = a^3 + 3a^2b + 3ab^2 + b^3$—but what is the product of four lengths a, b, c, and d? To the Greek way of thinking there was no such thing, because we cannot imagine four lines in mutually perpendicular directions.

Thus, the Greek interpretation of numbers as lengths and products as rectangles or boxes has its limitations. Nevertheless, it remained as the mental picture of products long after the Greek concept of length was replaced by a general concept of number (see the next section for more on this development). For example, here is a passage from Newton (1665) in which even the product of whole numbers is described as their "rectangle":

> For ye number of points in wch two lines may intersect can never bee greater y^n ye rectangle of ye number of their dimensions.

Here the "lines" are what we would call algebraic curves, and their "dimensions" are their degrees, which are whole numbers. (Also, it should probably be pointed out that the "y" in Newton's time is "th" in modern English.) Finally, as late as 1863, the great number theorist Dirichlet appealed to the rectangle picture in order to explain why $ab = ba$ for whole numbers a and b. On page 1 of his *Lectures on Number Theory* he asks the reader to imagine objects arranged in a rows of b objects, or in b rows of a objects, and to realize that the number of objects is the same in each case.

Surely, nothing could be clearer. Nevertheless, it is surprising that the same idea applies to two vastly different kinds of quantity: lengths, which vary *continuously*, and whole numbers, which vary *discretely*, or in jumps. Finding a concept of number that embraces these two extremes is a long journey, which results in a new understanding and appreciation of the law $ab = ba$. It will take two chapters to complete, and the remainder of this chapter outlines the obstacles that have to be overcome.

Exercises

1.1.1   Give pictorial versions of the *distributive law* $a(b + c) = ab + ac$, and the identity $a^2 - b^2 = (a - b)(a + b)$.

1.1.2   Also explain why $(a + b)^3 = a^3 + 3a^2b + 3ab^2 + b^3$, with a picture of a suitable cube.

1.2   What Are Numbers?

Numbers answer two subtly different questions: how many and how much? the first is the simpler question, answered by the *natural numbers*

$$0, \quad 1, \quad 2, \quad 3, \quad 4, \quad 5, \quad 6, \quad 7, \quad \dots .$$

The natural numbers originated for the simple purpose of counting, but they somehow developed an intricate structure, with operations of addition and multiplication and (partially) subtraction and division. The subtraction operation invites an extension of the natural numbers to the *integers*,

$$\dots, \quad -3, \quad -2, \quad -1, \quad 0, \quad 1, \quad 2, \quad 3, \quad \dots$$

so that subtraction becomes fully defined. And the division operation invites an extension of the integers to the *rational numbers* m/n for all integers m and $n \neq 0$, so that division is defined for all nonzero rational numbers.

You will know the rules for operating on natural numbers, integers and rational numbers from elementary school but, almost certainly, no underlying reason for the rules will have been given. In Sect. 2.2 we will show that all the rules for operating on numbers stem from their original purpose of counting, whereby all natural numbers originate from 0 by repeatedly adding 1.

You will also know from school that the rational numbers give an approximate answer to the second question: how much? This is because quantities such as length, area, mass, and so on, can be measured to arbitrary precision by rational numbers. Indeed, we can measure to arbitrary precision by *finite decimals*, or *decimal fractions*, which are rational numbers of the form $m/10^n$, where m and n are integers. (For example, $3.14 = 314/10^2$.) But arbitrary precision is not exactness, and some quantities are not exactly equal to any rational number.

The most famous example is the length, $\sqrt{2}$, of the diagonal of the unit square. The best-known proof goes as follows.

Irrationality of $\sqrt{2}$. *There is no rational number whose square equals* 2.

Proof. Suppose on the contrary that $2 = m^2/n^2$ for some positive integers m and n. Then we have the following series of implications.

$$2 = m^2/n^2 \Rightarrow 2n^2 = m^2 \qquad\qquad \text{(multiplying both sides by } n^2\text{)}$$

$$\Rightarrow m^2 \text{ is even}$$

$$\Rightarrow m \text{ is even} \qquad \text{(because the square of an odd number is odd)}$$

$$\Rightarrow m = 2m' \qquad\qquad\qquad \text{(for some natural number } m'\text{)}$$

$$\Rightarrow 2n^2 = (2m')^2 \qquad\qquad \text{(substituting } 2m' \text{ for } m \text{ in second line)}$$

$$\Rightarrow n^2 = 2m'^2 \qquad \text{(dividing both sides by 2)}$$

$$\Rightarrow n \text{ is even} \qquad \text{(by same argument as used above for } m)$$

$$\Rightarrow n = 2n' \qquad \text{(for some natural number } n')$$

$$\Rightarrow 2 = m'^2/n'^2 \qquad \text{(where } m' < m \text{ and } n' < n).$$

So, for any pair m, n of natural numbers with $2 = m^2/n^2$ there is a *smaller* pair m', n' with the same property, and we therefore have an infinite descending sequence of natural numbers, which is impossible. □

Thus, geometry demands *irrational* numbers. This discovery threw arithmetic into confusion, because it is not clear how to add and multiply irrational numbers. For example, is it true that $\sqrt{2} \times \sqrt{3} = \sqrt{6}$? Also, is there an arithmetic definition of multiplication compatible with geometry, where $\sqrt{2} \times \sqrt{3}$ measures the area of the rectangle with adjacent sides $\sqrt{2}$ and $\sqrt{3}$? We take up these questions in Sect. 2.4.

Exercises

If one attempts to prove that $\sqrt{3}$ is irrational by supposing that $\sqrt{3} = m/n$ and reasoning as above, one reaches the equation $n^2 = 3m^2$. It no longer follows that n^2 is even.

1.2.1   What property of n^2 *does* follow from the equation $n^2 = 3m^2$?
1.2.2   Use this property to devise a proof that $\sqrt{3}$ is irrational.
1.2.3   Also give a proof that $\sqrt{5}$ is irrational.

1.3 What Is the Line?

More precisely, what are *points*, and how do they fill the line? Or, how do we make a *continuum* from points? We would like to say that points on the line are numbers, but it is hard to recreate the uniform and unbroken quality of the line from our fragmentary perception of individual numbers. It is possible, certainly, to visualize the integer points on the line (Fig. 1.4)

Extending this vision to all the rational points is already a challenging task, because the rational points are *dense*—i.e., there are infinitely many of them in any interval of the line, no matter how small. One way to cope with density is to consider the integer points $\langle m, n \rangle$ of the *plane*, and to view the rational numbers

Fig. 1.4 Integer points on the line

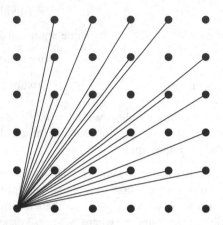

Fig. 1.5 Slopes to integer points on the plane

as the (slopes n/m of) lines from $\langle 0, 0 \rangle$ to other integer points $\langle m, n \rangle$ with $m \neq 0$.[1] (Figure 1.5 shows some of the positive rational numbers as slopes of lines from the origin to integer points in the first quadrant.)

This view also includes the *irrational* numbers in the form of lines through $\langle 0, 0 \rangle$ that *miss* all other integer points in the plane.

However, while it is nice to visualize the density of the rationals, and the gaps in them that correspond to irrationals, this picture brings us no closer to an *arithmetic* definition of the points on the line. For this, we need infinite concepts of some kind, so as to approach the irrational points via the rationals. The most familiar is probably the concept of the *infinite decimal*, which embraces both rational and irrational points in a uniform way. Infinite decimals extend finite decimals in a natural way and, like the finite decimals, they have a clear *ordering* corresponding to the ordering of points on the line.

As we said in Sect. 1.2, a finite decimal is a rational number of the form $m/10^n$, which we write by inserting a *decimal point* before the last n digits of the decimal numeral for m. Thus, 3.14 is the decimal form of $314/10^2$. An *infinite decimal*, such as

$$1.414213\ldots,$$

---

[1] In this book we use $\langle$ and $\rangle$ to bracket ordered pairs, triples, and so on. This is to avoid confusion with the notation (a, b), which will later be used for the *open interval* of points x such that $a < x < b$.

represents the *limit* of the finite decimals

$$1.4, \quad 1.41, \quad 1.414, \quad 1.4142, \quad 1.41421, \quad 1.414213, \quad \ldots$$

that is, the number "approached" by these finite decimals as the number of decimal places increases. We will define the concept of limit formally in Sect. 2.6, and simply assume some familiarity with infinite decimals for the present.

It is intuitively plausible that each point on the line corresponds to an infinite decimal, because we can find the successive decimal places of any point P by starting with the integer interval containing P, dividing the interval into 10 equal parts to find the first decimal place of P, dividing that subinterval into 10 equal parts to find the second decimal place, and so on. It is also plausible that different points P and Q will have different decimals, because repeated subdivision will eventually "separate" P from Q—they will eventually fall within different parts of the nth subdivision, and hence differ in the nth decimal place. Moreover, we can decide which of P, Q is less from their decimals—the lesser point is the one with the lesser digit in the first decimal place where they differ.

Thus, infinite decimals give a simple numerical representation of points on the line, which is particularly convenient for describing the *ordering* of points. However, infinite decimals are *not* convenient for describing addition and multiplication, so they are not a useful solution of the problem of defining addition and multiplication of irrational numbers. We will solve the latter problem differently in Sect. 2.4.

Exercises

Infinite decimals are also good for distinguishing (theoretically) between rational and irrational numbers: *the rational numbers are those with ultimately periodic decimals*. To see why any ultimately periodic decimal represents a rational number one uses an easy computation with decimals; namely, multiplication by 10 (possibly repeated), which shifts all digits one place to the left, and subtraction.

1.3.1 If $x = 0.37373737\ldots$, express x as a ratio of integers.
1.3.2 If $y = 0.519191919\ldots$, express y as a ratio of integers.
1.3.3 By generalizing the idea of the previous exercises, explain why each ultimately periodic decimal represents a rational number.
1.3.4 Find the decimals for 1/6 and 1/7.
1.3.5 By means of the division process, or otherwise, explain why each rational number has an ultimately periodic decimal.

The picture of integer points in the plane, used above to visualize rational numbers, has an interesting extension.

1.3.6 If each rational point in the plane is surrounded by a disk of fixed size ε, show that there is *no* line from $\langle 0, 0 \rangle$ that misses all other disks.
1.3.7 Conclude that, if space were filled uniformly with stars of uniform size, the whole sky would be filled with light (the Olbers paradox).

1.4 What Is Geometry?

In a trivial sense, numbers are related to geometry because the real numbers are motivated by our mental image of the line. However, the geometry of the line is not very interesting, compared with the geometry of the plane. What properties of numbers, if any, are relevant to the geometry of the plane? The short answer is: *Pythagorean triples.*

These are whole number triples $\langle a, b, c \rangle$ such that $a^2 + b^2 = c^2$ or, equivalently, pairs of whole numbers $\langle b, c \rangle$ such that $c^2 - b^2$ is a perfect square. A list of such pairs occurs in clay tablet known as Plimpton 322, which was inscribed around 1800 BCE in ancient Mesopotamia. Table 1.1 shows these pairs in modern notation, together with the number $a = \sqrt{c^2 - b^2}$, and a fraction x that will be explained later.

The original tablet lists the pairs $\langle b, c \rangle$ in the order given above (part of it is broken off) but not the numbers a, without which the numbers b and c look almost random and meaningless. The first person to realize that the pairs $\langle b, c \rangle$ are mathematically significant was the mathematics historian Otto Neugebauer, who in 1945 noticed that $\sqrt{c^2 - b^2}$ is a whole number in each case. [See Neugebauer and Sachs (1945).] This led him to suspect that Plimpton 322 was really a table of triples $\langle a, b, c \rangle$ with the property

$$a^2 + b^2 = c^2.$$

Such triples are called Pythagorean because of the famous *Pythagorean theorem* asserting that $a^2 + b^2 = c^2$ holds in any right-angled triangle with sides a, b and hypotenuse c. It can hardly be a coincidence that the numbers b, c have the *numerical* property that $\sqrt{c^2 - b^2}$ is a whole number, but was the compiler of Plimpton 322

Table 1.1 Numbers in Plimpton 322, and related numbers

a	b	c	x
120	119	169	12/5
3,456	3,367	4,825	64/27
4,800	4,601	6,649	75/32
13,500	12,709	18,541	125/54
72	65	97	9/4
360	319	481	20/9
2,700	2,291	3,541	54/25
960	799	1,249	32/15
600	481	769	25/12
6,480	4,961	8,161	81/40
60	45	75	2
2,400	1,679	2,929	48/25
240	161	289	15/8
2,700	1,771	3,229	50/27
90	56	106	9/5

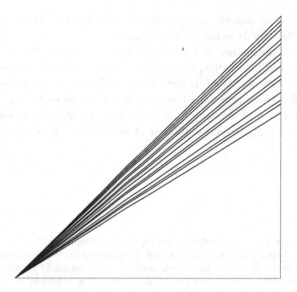

Fig. 1.6 Triangle shapes from Plimpton 322

thinking about triangles? What makes this virtually certain is that the ratio b/a (the "slope of the hypotenuse") decreases steadily, and in roughly equal steps from just below the 45° slope to just above the 30° slope. Thus, the triples are geometrically ordered, and geometrically bounded, in a natural way. Moreover, it has been pointed out by Christopher Zeeman (see exercises below) that there are just 16 triangles between these bounds, subject to a certain "simplicity" condition, and Plimpton 322 contains the first 15 of them. Figure 1.6 shows the shapes of the 15 triangles in question.

Thus, Neugebauer's discovery suggests that the numbers in Plimpton 322 have a *geometric meaning*, and that the Pythagorean theorem was known long before Pythagoras, who lived around 500 BCE.

The ordering of triples $\langle a, b, c \rangle$ by the ratios b/a makes it clear that *positive rational numbers* (ratios of positive integers) were also part of ancient mathematical thinking. So it is reasonable to suppose that rational numbers may have been involved in the generation of the pairs $\langle b, c \rangle$ in Plimpton 322, and that may explain how huge pairs such as $\langle 12709, 18541 \rangle$ could be discovered. One way to do this is via the fractions x appended in the last column of the table. The fractions x "explain" the triples $\langle a, b, c \rangle$ in the sense that they yield each a, b and c by the formulas

$$\frac{b}{a} = \frac{1}{2}\left(x - \frac{1}{x}\right), \quad \frac{c}{a} = \frac{1}{2}\left(x + \frac{1}{x}\right)$$

and each x is considerably shorter than the triple it generates. Moreover, the fractions x and $1/x$ are noteworthy because each has a *finite expansion* in base 60, the number system used in ancient Mesopotamia. This is because the numerator and denominator of each factorizes into powers of 2, 3, or 5. These properties are explored more thoroughly in the exercises below.

Like the Mesopotamians, the Pythagoreans were struck by the presence of whole numbers, and their ratios, in geometry (and elsewhere, particularly music). However, as we saw in Sect. 1.2, they discovered *gaps* in the rational numbers, which makes the whole program of using numbers in geometry problematic. The gap at $\sqrt{2}$ is also conspicuous by its absence from Plimpton 322: the hypotenuse lines of slope b/a stop just short of the line of slope 1 corresponding to the diagonal of a square.

Exercises

We now explore how each Pythagorean triple $\langle a, b, c \rangle$ in Plimpton 322 is "explained" by the fraction x in the last column of Table 1.1. Notice that x generally has the same number of digits as each of the numbers a, b, c, so all three of these numbers can be encoded by a number of the same "length" as any one of them. Also, we will see that in each case x is "simple" in a certain sense.

For each line in the table,

$$\frac{b}{a} = \frac{1}{2}\left(x - \frac{1}{x}\right).$$

1.4.1   Check that $\frac{1}{2}\left(x - \frac{1}{x}\right) = \frac{119}{120}$ when $x = 12/5$.
1.4.2   Also check, for three other lines in the table, that

$$\frac{b}{a} = \frac{1}{2}\left(x - \frac{1}{x}\right).$$

The numbers x are not only "shorter" than the numbers b/a, they are "simple" in the sense that they are built from the numbers 2, 3, and 5. For example

$$\frac{12}{5} = \frac{2^2 \times 3}{5} \quad \text{and} \quad \frac{125}{54} = \frac{5^3}{2 \times 3^3} \ .$$

1.4.3   Check that every other fraction x in the table can be written with both numerator and denominator as a product of powers of 2, 3, or 5.

The formula $\frac{1}{2}\left(x - \frac{1}{x}\right) = \frac{b}{a}$ gives us whole numbers a and b from a number x. Why should there be a whole number c such that $a^2 + b^2 = c^2$?

1.4.4   Verify by algebra that $\left[\frac{1}{2}\left(x - \frac{1}{x}\right)\right]^2 + 1 = \left[\frac{1}{2}\left(x + \frac{1}{x}\right)\right]^2$.
1.4.5   Deduce from Exercise 1.4.4 that $a^2 + b^2 = c^2$, where $\frac{1}{2}\left(x + \frac{1}{x}\right) = \frac{c}{a}$.
1.4.6   Check that the formula in Exercise 1.4.5 gives $c = 169$ when $x = 12/5$ (the first line of the table), and also check three other lines in the table.

In a 1995 talk at the University of Texas at San Antonio, Christopher Zeeman showed that there are exactly 16 fractions x with denominator less than 60 and numerator and denominator composed of factors 2, 3, and 5 that give slope b/a corresponding to an angle between 30° and 45°. The Pythagorean triples in Plimpton 322 correspond to the first 15 slopes (in decreasing order) obtainable from these values of x.

1.4.7 The slope not included in Plimpton 322 comes from the value $x = 16/9$. Show that the corresponding slope is 175/288, and that the corresponding angle is a little more than 30°.

1.5 What Are Functions?

Having seen how the rational numbers can be "completed" to the line of all real numbers, we might hope that there are similar "rational functions" among all the real functions. There are indeed rational functions. They are the functions that are quotients of *polynomials*

$$p(x) = a_0 + a_1 x + a_2 x^2 + \cdots + a_n x^n, \quad \text{where } a_0, a_1, a_2, \ldots, a_n \text{ are real numbers.}$$

The polynomial functions play the role of integers, and they are sometimes called "integral rational functions."

Just as rational numbers fail to exhaust all numbers, rational functions fail to exhaust all functions, and indeed there are many naturally occurring functions that are not rational, such as $\sin x$ and $\cos x$. We can see that the function $\sin x$ is not rational, because it is zero for infinitely many values of x—namely, $x = 0, \pm\pi, \pm 2\pi, \pm 3\pi, \ldots$. A rational function, on the other hand, is zero only when its numerator is zero. This happens for at most n values of x, where n is the degree of the numerator, by the fundamental theorem of algebra.

The functions $\sin x$ and $\cos x$ are nevertheless *limits* of rational functions, and indeed of polynomials, because

$$\sin x = x - \frac{x^3}{3!} + \frac{x^5}{5!} - \frac{x^7}{7!} + \cdots,$$

$$\cos x = 1 - \frac{x^2}{2!} + \frac{x^4}{4!} - \frac{x^6}{6!} + \cdots.$$

Thus, $\sin x$ is the limit of the sequence of polynomials

$$x, \quad x - \frac{x^3}{3!}, \quad x - \frac{x^3}{3!} + \frac{x^5}{5!}, \ldots,$$

just as $\sqrt{2}$ is the limit of the sequence of rationals $1, 1.4, 1.41, 1.414, \ldots$. The infinite series that occur as limits of polynomials are called *power series*, and they form an important class of functions. It is tempting to think that power series "complete" the rational functions in the same way that the real numbers complete the rational numbers, and indeed Newton was extremely impressed by the analogy:

> I am amazed that it has occurred to no one ... to fit the doctrine recently established for decimal numbers in similar fashion to variables, especially since the way is then open to

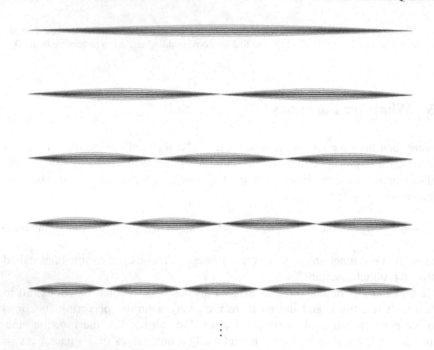

Fig. 1.7 Modes of vibration

more striking consequences. For since this doctrine in species has the same relationship
to Algebra that the doctrine in decimal numbers has to common Arithmetic, its operations
of Addition, Subtraction, Multiplication, Division and Root extraction may be easily learnt
from the latter's.

                                                                    Newton (1671), p. 35.

Power series do indeed vastly increase the range of functions to which algebraic
operations apply, and they are also subject to the calculus operations of differentiation and integration, as Newton was well aware. Nevertheless, power series by
no means exhaust all possible functions. A much wider class of functions came to
light in the eighteenth century when mathematicians investigated the problem of the
vibrating string.

A taut string with two fixed ends has many *simple modes of vibration*, the first
few of which are shown in Fig. 1.7.

The shape of the first mode is one-half of the sine curve $y = \sin x$ (scaled down
in the y dimension), the shape of the second is $y = \sin 2x$, and so on. These simple
modes correspond to simple *tones*, of higher and higher pitch as the number of
waves increases, and they may be summed to form compound tones corresponding
to (possibly infinite) sums

$$b_1 \sin x + b_2 \sin 2x + b_3 \sin 3x + \cdots$$

Fig. 1.8 Triangular wave

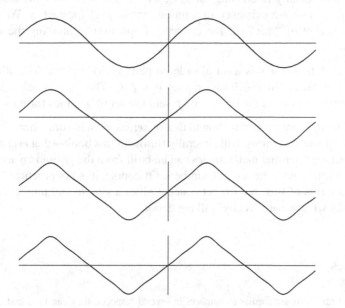

Fig. 1.9 Approximations to the triangle wave

It seems, conversely, that an arbitrary continuous wave form may be realized by such a sum. This was first conjectured by Bernoulli (1753), and his remark led to the realization that trigonometric series were even more arbitrary than power series. For example, a power series may be differentiated at any point, so the corresponding curve has a tangent at any point. This is not the case for the "triangular wave" shown in Fig. 1.8, which has no tangent at its highest points.

However, the triangular wave is the sum of the infinite series

$$\sin x - \frac{\sin 3x}{9} + \frac{\sin 5x}{25} - \frac{\sin 7x}{49} + \cdots$$

Figure 1.9 shows the sums of the first one, two, three, and four terms of this series, and how they approach the triangular wave shape. This discovery made it acceptable to consider any continuous graph (on a finite interval) as the graph of a function, because one could express any such graph by a trigonometric series.

Indeed, why insist that "functions" be continuous? Why not allow the following rule to define a function? (It is known as the *Dirichlet function*.)

$$d(x) = \begin{cases} 1 \text{ if } x \text{ is rational} \\ 0 \text{ if } x \text{ is irrational.} \end{cases}$$

This rule defines a unique value of $d(x)$ for each x, which is perhaps all we need ask of a function. Interesting though it may be to find formulas for functions (and indeed it is especially interesting for $d(x)$, as we will see later), the *essence* of a function f is the mere existence of a unique value $f(x)$ for each x. We can even strip the concept of "rule" off the concept of function by making the following definition.

Definition. A *function* f is a set of ordered pairs $\langle x, y \rangle$ that includes at most one pair $\langle x, y \rangle$ for each x, in which case we say $y = f(x)$. The set of x values occurring in the ordered pairs form the *domain* of f, and the set of y values form its *range*.

Reducing the concept of function to that of *set*, as we have done here, is part of a view of mathematics that we will generally follow in this book—that *everything is a set*. We will see later that mathematics can be built from the ground up according to this view, starting with the natural numbers. Of course, it is not practical constantly to think in terms of sets, but the set concept gives a simple and uniform answer if anyone asks what we are "really" talking about.

Exercises

The rational functions are similar to numbers in several respects: they can be added, subtracted, multiplied, and divided (except by 0), and they satisfy the same rules of algebra, such as $a+b = b+a$ and $ab = ba$. More surprisingly, they can be *ordered*. If f and g are rational functions, and $f \neq g$, we say that $f < g$ or $g > f$ if $g(x)$ is *ultimately* greater than $f(x)$; that is, if $g(x) > f(x)$ for all sufficiently large x. Then, *if f and g are any rational functions, either $f \leq g$ or $g \leq f$.* The following exercises explain why.

1.5.1 If $f(x) = x^2$, $g(x) = x^2 + 100$, $h(x) = x^3$, show that $f < g < h$.

1.5.2 If f and g are polynomials, explain how to tell which of f, g is the greater.

1.5.3 Show that $\frac{x+1}{x-1} > \frac{x-1}{x+1}$ for x sufficiently large.

1.5.4 If f and g are rational functions, explain how to tell which of f, g is greater by reducing to the same question about polynomials.

Notice also that the rational functions include a copy of the real numbers, if we associate each real number a with the *constant* function $f(x) = a$. Thus, the ordered set of rational functions is a kind of "expanded line," with new points corresponding to the nonconstant functions.

The new points, however, make the set of rational functions less suitable, as a model of the intuitive line, than the set of real numbers. One reason is that the rational function "line" includes *infinitesimals*—positive functions that are smaller than any positive real number.

1.5.5 Show that the rational function $\iota(x) = 1/x$ is greater than zero but less than any constant function. (The symbol ι is the Greek letter iota, which seems appropriate for an infinitesimal function.)

1.6 What Is Continuity?

In the previous section we observed the concept of continuous function on our way
to the general concept of function. But the concept of continuous function, though
it arose earlier than the general concept, turns out to be much harder to define
precisely.

Our intuitive concept of a continuous function is one whose graph is an *unbroken*
curve. We note in passing that the concept of a "curve" is thereby linked to the
concept of a continuous function, and we later (Sect. 4.4) take up what this says
about the concept of curve. But for now we will concentrate on the meaning of
"unbrokenness," or the absence of gaps, which is a concept called *connectedness*.
We have already noted that connectedness is an attribute of the line. But the concept
of *continuous function* is more general, because it has meaning whether or not the
domain of the function is connected.

We will eventually want to study continuous functions on disconnected domains,
but for the present we restrict ourselves to functions on $\mathbb{R}$ or on intervals of $\mathbb{R}$ such
as $[0, 1] = \{x : 0 \le x \le 1\}$ or $(0, \infty) = \{x : x > 0\}$. In this case we can define a
function f to be *continuous* on the interval if, for any number a in the interval, $f(x)$
approaches $f(a)$ as x approaches a. There are various ways to formalize the idea of
"approaching," which we will discuss in later chapters. For now, we just illustrate
the idea with two examples of functions that visibly *fail* to be continuous at a point
$x = a$. The first example is the function

$$g(x) = \begin{cases} -1 & \text{for } x < 0 \\ 1 & \text{for } x \ge 0. \end{cases}$$

This function fails to be continuous at $x = 0$ because $g(x)$ approaches -1 (in fact
stays constantly equal to -1) as x approaches 0 from below, yet $g(0) = 1$. The
second example is the function

$$h(x) = \begin{cases} 0 & \text{for } x \le 0 \\ \sin \frac{1}{x} & \text{for } x > 0, \end{cases}$$

which fails to approach *any* value as x approaches 0 from above, because it oscillates
between -1 and 1 no matter how close x comes to 0. The graphs of g and h in
Fig. 1.10 clearly show the points of discontinuity.

$y = g(x)$

$y = h(x)$

Fig. 1.10 Graphs of the functions g and h

A continuous function on a closed interval $[a, b] = \{x : a \le x \le b\}$ has properties that one would expect from the intuition that its graph is an unbroken curve from the point $\langle a, f(a) \rangle$ to the point $\langle b, f(b) \rangle$. Notably, the following:

Intermediate Value Property. *If f is a continuous function on the closed interval $[a, b]$, then f takes each value between $f(a)$ and $f(b)$.*

Extreme Value Property. *If f is a continuous function on the closed interval $[a, b]$, then f takes a maximum and a minimum value on $[a, b]$.*

As obvious as these properties appear, they are far from trivial to prove, and they depend on the connectedness property of $\mathbb{R}$. Indeed these properties are logically and historically the reason why we need to study the real numbers in the first place. The need to prove the intermediate value property became pressing after Gauss (1816) used it in a proof of the *fundamental theorem of algebra*, which states that each polynomial equation has a solution in the complex numbers. The first attempt to prove the intermediate value property was made by Bolzano (1817), but Bolzano's proof rests on another assumption that one would like to be provable— the *least upper bound property* of $\mathbb{R}$, according to which any bounded set of real numbers has a least upper bound. In 1858, Dedekind first realized that the least upper bound property is rigorously provable from a suitable definition of $\mathbb{R}$, thus providing a sound foundation for all the basic theorems of calculus. This is why whole books have been written about $\mathbb{R}$, such as Dedekind (1872), Huntington (1917), and Landau (1951). We explain the construction of $\mathbb{R}$, and prove its basic properties, in Chap. 2.

But the study of functions also demands that we study *sets* of real numbers. For example, when a function is discontinuous we may wish to understand how far it departs from continuity, and this involves studying the set of points where it is discontinuous. A basic question then is how large (or small) a set of points may be, and hence: how can we *measure* a set of real numbers? We expand on this question in the next section.

Exercises

1.6.1 Show that $x^k - a^k = (x - a)(x^{k-1} + ax^{k-2} + a^2 x^{k-3} + \cdots + a^{k-1})$.

1.6.2 Deduce from Exercise 1.6.1 that, $x - a$ divides $p(x) - p(a)$ for any polynomial $p(x)$.

1.6.3 Deduce from Exercise 1.6.2 that each root $x = a$ of a polynomial equation $p(x) = 0$ corresponds to a factor $(x - a)$ of $p(x)$, and hence that the equation $p(x) = 0$ has at most n roots when p has degree n.

1.6.4 Show, using the intermediate value property, that $x^5 + x + 1 = 0$ has at least one real root.

1.6.5 More generally, show that a polynomial equation $p(x) = 0$ has a real root for any polynomial p of odd degree.

1.7   What Is Measure?

It is intuitively plausible that every set S of real numbers, say in [0,1], should have a measure, because it seems meaningful to ask: if we choose a point x at random in [0,1], what is the probability that x lies in S? This probability, if it exists, can be taken as the measure of S. Certain simple sets certainly have a measure in the sense of this thought experiment. The measure of the interval $[a, b]$ should be $b - a$, and this should also be the measure of (a, b), because the measure of the point a or b should be zero.

A more interesting set is the set of rationals in [0,1]. If we ask what is the probability that a random point is rational, the surprising answer is zero! This is because we can cover all the rational numbers in [0,1] by intervals of total length $\leq \varepsilon$, *for any number $\varepsilon > 0$*. To see why, notice that we can arrange all the rationals in [0,1] in the following list (according to increasing denominator):

$$0, \quad 1, \quad \frac{1}{2}, \quad \frac{1}{3}, \frac{2}{3}, \quad \frac{1}{4}, \frac{3}{4}, \quad \frac{1}{5}, \frac{2}{5}, \frac{3}{5}, \frac{4}{5}, \quad \frac{1}{6}, \frac{5}{6}, \quad \cdots$$

So if we cover the nth rational on the list by an interval of length $\varepsilon/2^{n+1}$ then

$$\text{total length covered} \leq \frac{\varepsilon}{2} + \frac{\varepsilon}{4} + \frac{\varepsilon}{8} + \cdots = \varepsilon.$$

Since ε can be made as small as we please, the total measure of the rational numbers can only be zero. This is surely surprising! In fact, we have exposed two surprising facts:

1. The rational numbers in [0,1] can be arranged in a list.
2. Any listable set has measure zero, and hence it is not the interval [0,1].

These two facts together give another proof that irrational numbers exist; in fact they show that "almost all" numbers are irrational, because the probability that a randomly chosen number is rational is zero.

Thus, the concept of measurability leads to unexpected discoveries about sets of real numbers. Even more surprising results come to light as we pursue the concept further, as we will do in later chapters.

1.7.1   Area and Volume

The idea of determining measure by adding infinitely many items actually goes back to ancient Greece, where the method was used by Euclid and Archimedes to find certain areas and volumes. A spectacular example was Archimedes' determination of the *area of a parabolic segment*, which he found by filling the parabolic segment by infinitely many triangles. His method is illustrated in Fig. 1.12. For simplicity

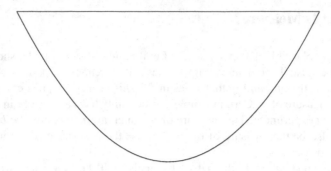

Fig. 1.11 The parabolic segment

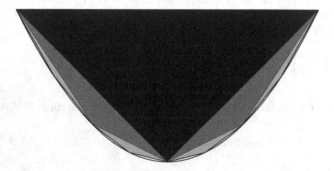

Fig. 1.12 Filling the parabolic segment with triangles

we take the parabola $y = x^2$, and cut off the segment shown in Fig. 1.11, between $x = -1$ and $x = 1$.

The first triangle has vertices at the ends and midpoint of the parabola. Between each of its lower edges and the parabola we insert a new triangle whose third vertex is also on the parabola, halfway (in x value) between the other two. Then we repeat the process with the lower edges of the new triangles. This creates successive "generations" of triangles, with each triangle in generation $n + 1$ having two vertices from a triangle in generation n and its third vertex also on the parabola halfway (in x value) between them. The first three generations are shown in black, gray, and light gray in Fig. 1.12.

It is easy to check (see exercises) that the area of generation $n + 1$ is 1/4 of the area of generation n, so we can find the area by summing a geometric series:

area of parabolic segment

$$= \left(1 + \frac{1}{4} + \frac{1}{4^2} + \frac{1}{4^3} + \cdots\right) \times \text{area of the first triangle} = \frac{4}{3} \times 1 = \frac{4}{3}.$$

Exercises

The following exercises verify the areas of the triangles in Archimedes' construction.

1.7.1 Show that the triangle with vertices (which lie on the parabola $y = x^2$)

$$\langle a, a^2 \rangle, \left\langle \frac{a+b}{2}, \left(\frac{a+b}{2}\right)^2 \right\rangle, \langle b, b^2 \rangle \quad \text{has area} \quad \left(\frac{b-a}{2}\right)^3.$$

1.7.2 Deduce from Question 1.7.1 that each triangle in generation n has area 2^{3-3n}, and hence that the total area of generation n is 2^{2-2n}.

1.7.3 Deduce from Question 1.7.2 that the total area of all triangles inside the parabolic segment is 4/3.

Virtually the same geometric series occurs in Euclid's determination of the volume of the tetrahedron, in the *Elements*, Book XII, Proposition 4. Euclid uses two prisms whose edges join the midpoints of the tetrahedron edges (Fig. 1.13).

1.7.4 Assuming that the volume of a prism is triangular base area × height, show that the prisms in Fig. 1.13 have volume equal to 1/4 (tetrahedron base area) × (tetrahedron height).

Now remove these two prisms, and repeat the process in the two tetrahedra that remain, as in Fig. 1.14.

1.7.5 By repeating the argument of Exercise 1.7.4, show that

$$\text{volume of tetrahedron} = \left(\frac{1}{4} + \frac{1}{4^2} + \frac{1}{4^3} + \cdots\right) \text{base area} \times \text{height}$$

$$= 1/3 \text{ base area} \times \text{height}.$$

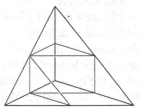

Fig. 1.13 Euclid's prisms inside a tetrahedron

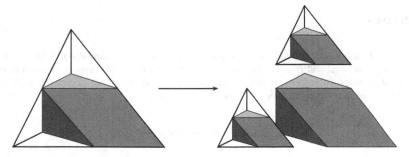

Fig. 1.14 Repeated dissection of the tetrahedron

1.8   What Does Analysis Want from $\mathbb{R}$?

From the discussions in the preceding sections, we expect that answers to several fundamental questions about numbers, functions, and curves will emerge from a better understanding of the system $\mathbb{R}$ of real numbers. To obtain good answers, we seem to want $\mathbb{R}$ to have the following (as yet only vaguely defined) properties.

1. Algebraic structure.

     Since the members of $\mathbb{R}$ are supposed to be numbers, they should admit sum, difference, product, and quotient operations, subject to the usual rules of algebra. For example, it should be true that $ab = ba$ and that $(\sqrt{2})^2 = 2$.
2. Completeness.

     $\mathbb{R}$ should be *arithmetically complete*, in the sense that certain infinite operations on $\mathbb{R}$ have results in $\mathbb{R}$. For example, the infinite sum represented by an infinite decimal, such as

$$3.14159\cdots = 3 + \frac{1}{10} + \frac{4}{10^2} + \frac{1}{10^3} + \frac{5}{10^4} + \frac{9}{10^5} + \cdots$$

should be a member of $\mathbb{R}$.

     $\mathbb{R}$ should also be *geometrically complete*, in the sense of having no gaps, like a continuous line. From this property we hope to derive a concept of *continuous function* with the expected properties, such as the intermediate value property.

     Hopefully, arithmetic and geometric completeness will be equivalent, so both can be achieved simultaneously. Also, if $\mathbb{R}$ behaves like a line, then $\mathbb{R}^2$ should behave like a plane.
3. Measurability of subsets (of both $\mathbb{R}$ and $\mathbb{R}^2$).

     We hope for a definition of measure that gives a definite measure to each subset of $\mathbb{R}$, or at least to each "clearly defined" subset of $\mathbb{R}$. We hope the same for subsets of $\mathbb{R}^2$, because the subsets of the plane $\mathbb{R}^2$ include the "regions under curves $y = f(x)$," the measure of which should represent the *integral* of f.

     Thus, the problem of finding the measure of plane sets includes the problem of finding integrals—one of the fundamental problems of analysis.

Exercises

Large parts of analysis also depend on the *complex* numbers, which are numbers of the form $a + bi$, where a and b are real and $i^2 = -1$. To avoid a separate discussion of complex numbers in this book we show that their properties reduce to those of the real numbers. An elegant way to do this is to represent each complex number $a + bi$ by the 2×2 matrix

$$\begin{pmatrix} a & b \\ -b & a \end{pmatrix} \quad \text{where } a \text{ and } b \text{ are real.}$$

Then the sum and product of complex numbers are the sum and product of matrices, which are defined in terms of the sum and product of real numbers, so they inherit their algebraic properties from those of the real numbers.

1.8.1 Writing

$$\begin{pmatrix} a & b \\ -b & a \end{pmatrix} = a \begin{pmatrix} 1 & 0 \\ 0 & 1 \end{pmatrix} + b \begin{pmatrix} 0 & 1 \\ -1 & 0 \end{pmatrix} = a\mathbf{1} + b\mathbf{i},$$

show that $\mathbf{i}^2 = -\mathbf{1}$.

1.8.2 If $\mathbf{0} = \begin{pmatrix} 0 & 0 \\ 0 & 0 \end{pmatrix}$, and $a\mathbf{1} + b\mathbf{i} \ne \mathbf{0}$, show that $(a\mathbf{1} + b\mathbf{i})^{-1}$ exists, and

$$(a\mathbf{1} + b\mathbf{i})^{-1} = \frac{1}{a^2 + b^2}(a\mathbf{1} - b\mathbf{i}).$$

1.8.3 Show, using the properties of matrix addition and multiplication, that for any complex numbers u, v, w:

$$u + v = v + u, \qquad uv = vu,$$
$$u + (v + w) = (u + v) + w, \qquad u(vw) = (uv)w,$$
$$u + \mathbf{0} = u, \qquad u \cdot \mathbf{1} = u,$$
$$u + (-u) = \mathbf{0}, \qquad u \cdot u^{-1} = \mathbf{1} \quad \text{for } u \ne \mathbf{0}$$
$$u(v + w) = uv + uw.$$

1.9 Historical Remarks

As can be seen from the early sections of this chapter, some fundamental problems of analysis arose long before the development of calculus, and they were not solved until long after. It is fair to say, however, that calculus focused the attention of mathematicians on infinite processes, and it drove the search for answers to the fundamental questions. It turned out that the ancients themselves were close to answers—or so it seems with the advantage of hindsight—but they were held back by fear of infinity.

Much of what we know about the ancient Greek understanding of numbers and geometry comes from Euclid's *Elements*, written around 300 BCE. We know that Euclid collected ideas from earlier mathematicians, such as Eudoxus, but the *Elements* is the first known systematic presentation of mathematics. It covers both geometry and number theory, and it struggles with the problem that divides them: the existence of irrational quantities. The longest book in the *Elements*, Book X, is devoted to the classification of irrational quantities arising in geometry. And the most subtle book in the *Elements*, Book V, is devoted to Eudoxus' "theory of proportions," which seeks to deal with irrational quantities by comparing them with the rationals.

In the next chapter we will see how that idea of comparing an irrational quantity with rationals was revived by Dedekind in the nineteenth century to provide a concept of *real numbers* making up a *number line*, thus providing an arithmetical foundation for geometry. The novel part of Dedekind's idea is its acceptance of infinite sets—an idea that the Greeks rejected.

Another idea of Eudoxus, the "method of exhaustion," was also pushed further in the late nineteenth century *theory of measure*. Just as the theory of proportions compares a complicated (irrational) number with simple ones (rationals), the method of exhaustion compares a complicated geometric object with simple (and measurable) ones, such as triangles or rectangles. A typical example of exhaustion is Archimedes' determination of the area of a parabolic segment by comparing it with collections of triangles (Sect. 1.7). Although there are infinitely many triangles in the construction, the Greeks avoided considering their infinite totality by showing that the area of the parabolic segment can be approximated arbitrarily closely by *finitely many* of them.

Thus, by summing a finite geometric series, one can show that any area less than 4/3 may be exceeded by a finite collection of triangles inside the parabolic segment. The possibility that the segment has area less than 4/3 is thereby ruled out, and the only remaining possibility is that its area equals 4/3. This is what the word "exhaustion" means in this context: one finds the exact value of the area by exhausting all other possible values. In the late nineteenth century it was found that extremely complicated geometric objects, called *measurable sets*, could be measured by a similar method. The objects in question are again approximated by finite collections of simple objects (line intervals or rectangles), but showing that the approximation is arbitrarily close may require the use of infinite collections. This will be explained in Chap. 9.

In this sense, we can say that the ancient Greeks came close to answering the basic questions about number, geometry, and measure. The questions about functions and continuity are another story, very much a product of the development of calculus in the seventeenth and eighteenth centuries.

As mentioned in Sect. 1.5, "functions" were originally things described by "formulas," though formulas could be infinite power series or trigonometric series. But when Bernoulli (1753) conjectured that the shape of an arbitrary string could be expressed by a trigonometric series it was still thought that such a function must be continuous. This was disproved by the general theory of trigonometric series developed by Fourier (1807). Among other examples, Fourier exhibited the "square wave" function

$$\cos x - \frac{\cos 3x}{3} + \frac{\cos 5x}{5} - \frac{\cos 7x}{7} + \cdots$$

which jumps from the value $\pi/4$ on the interval $(-\pi/2, \pi/2)$ to the value $-\pi/4$ on the interval $(\pi/2, 3\pi/2)$.

Despite such examples, Fourier tended to assume that functions defined by trigonometric series are continuous. Dirichlet (1837) was the first to insist that

Fig. 1.15   Daniel Bernoulli and Joseph Fourier

arbitrary functions really *can* have arbitrary values, and any general argument about functions should cover discontinuous functions. As an example, he introduced the function we call the *Dirichlet function*,

$$d(x) = \begin{cases} 1 \text{ if } x \text{ is rational} \\ 0 \text{ if } x \text{ is irrational.} \end{cases}$$

In fact, by the end of the nineteenth century, the Dirichlet function did not seem especially pathological. It is a limit of limits of continuous functions, and can be expressed by the formula

$$d(x) = \lim_{m \to \infty} \lim_{n \to \infty} (\cos(m!\pi x))^n$$

of Pringsheim (1899), p. 7.

Thus, the seventeenth-century idea of representing functions by formulas extends much further than was first thought. "Formulas" may not be available for all functions, but they extend far beyond the continuous functions—certainly to all functions obtainable from continuous functions by taking limits. These functions are called the *Baire functions* after René Baire, who first studied them in 1898. Baire functions, and their close relatives the *Borel sets*, will be discussed in Chap. 8.

The need to clarify the concept of continuity arose, as mentioned in Sect. 1.6, from attempts to prove the fundamental theorem of algebra, particularly the one by Gauss (1816). It should be added that not only did the solution come from outside algebra, via the intermediate value theorem of Bolzano (1817), so too did the problem itself. Originally, the motivation for a fundamental theorem of algebra was to integrate rational functions. The method of partial fractions makes it possible to integrate any quotient $p(x)/q(x)$ of polynomials, provided we can split $q(x)$ into real linear and quadratic factors. Such a factorization follows from the fundamental

Fig. 1.16 Carl Friedrich Gauss and Bernard Bolzano

theorem of algebra. The novel contribution of Gauss was to see that one should *not* attempt to find formulas for the roots of polynomial equations, but rather to deduce the *existence* of roots from general properties of continuous functions.

Thus, a problem about the most concrete kind of formulas, polynomials, was eventually solved by abstract reasoning about the general class of continuous functions. And, as Bolzano discovered, reasoning about continuous functions depends in turn on an abstract property of real numbers, the *least upper bound* property. It was to establish this property that Dedekind proposed his definition of the real numbers, which draws its inspiration from the ancient theory of proportions. Dedekind's remarkable fusion of ancient and modern ideas will be developed in the next chapter.

Chapter 2
From Discrete to Continuous

PREVIEW

The questions raised in the introductory chapter stem from a single problem: bridging the gap between the *discrete* and the *continuous*. Discreteness is exemplified by the positive integers $1, 2, 3, 4, \ldots$, which arise from counting but also admit addition and multiplication. Continuity is exemplified by the concept of distance on a line, which arises from measurement but also admits addition and multiplication. The problem is to find a concept of *real number* that embraces both counting and measurement, and satisfies the expected laws of addition and multiplication.

We begin by laying the simplest possible foundation for arithmetic on the positive integers, the principle of *induction*, which expresses the idea that every positive integer can be reached by starting at 1 and repeatedly adding 1. The corresponding method of *proof* by induction then allows us to prove the basic laws of arithmetic. From here it is only a short step to the arithmetic of positive *rational* numbers—the ratios m/n of positive integers m and n.

With the laws of arithmetic established on the foundation of induction we can concentrate on constructing the real numbers by filling the gaps in the rationals. This is the step that completes the transition from discrete to continuous, and to carry it out we need an infinite process of some kind. The *Dedekind cut* process is the one favored in this book, since it extends the laws of arithmetic from rational to real numbers in a natural way. However, we also discuss some other infinite processes commonly used to describe real numbers, such as infinite decimals and continued fractions.

2.1 Counting and Induction

The origins of mathematics are lost in prehistory, but it seems reasonable to suppose that mathematics began with counting, so the first mathematical objects encountered were the positive integers $1, 2, 3, 4, 5, \ldots$. These objects, generated by the seemingly

J. Stillwell, *The Real Numbers: An Introduction to Set Theory and Analysis*, Undergraduate Texts in Mathematics, DOI 10.1007/978-3-319-01577-4_2, © Springer International Publishing Switzerland 2013

simple process of starting with 1 and then going from each number to its successor, are not only infinitely numerous, but also infinitely rich in beauty and complexity. Evidence from Mesopotamia (Sect. 1.4) suggests that it was known as early as 1800 BCE that positive integers have remarkable properties involving addition and multiplication, and by 300 BCE systematic *proofs* of such properties were given in Euclid's *Elements*.

Among the methods of proof in Euclid, one sees an early form of what we now call *induction*. Induction reflects the fact that each positive integer n can be reached from 1 by repeatedly adding 1. So if we start with the number n we can take only a finite number of *downward* steps. We have seen one such "descent" argument already: the proof that $\sqrt{2}$ is irrational given in Sect. 1.2. Euclid uses the "descent" form of induction to prove two important results.

1. Each integer $n > 1$ has a prime divisor.

    Because if n is not itself prime, it factorizes into smaller positive integers, m_1 and n_1, to which the same argument applies. Since each step of the process produces smaller numbers, it must terminate—necessarily on a prime divisor of n.

2. The Euclidean algorithm terminates on any pair of positive integers a, b.

    The Euclidean algorithm, as Euclid himself described it, "repeatedly subtracts the lesser number from the greater." That is, we start with the pair $\langle a_1, b_1 \rangle = \langle a, b \rangle$, for which we can assume $a > b$, and successively form the pairs

$$\langle a_2, b_2 \rangle = \langle \max(b_1, a_1 - b_1), \min(b_1, a_1 - b_1) \rangle,$$

$$\langle a_3, b_3 \rangle = \langle \max(b_2, a_2 - b_2), \min(b_2, a_2 - b_2) \rangle, \ldots$$

until we get $a_n = b_n$. Since the algorithm produces a decreasing sequence of numbers, termination necessarily occurs.

These two results are foundation stones of number theory. The first shows the existence of prime factorization for any integer > 1, and the second shows (after some other steps we omit here) the *uniqueness* of prime factorization. So induction is evidently a fundamental principle of proof in number theory.

Today, we know that virtually all of number theory can be encapsulated by a small set of axioms—the *Peano axioms*—that state the basic properties of the successor function, definitions of addition and multiplication, and induction. In fact, induction is implicit in the definitions of addition and multiplication, and in Sect. 2.2 we will see how their properties unfold when induction is applied. However, this was quite a late development in the history of mathematics. It took a long time for the idea of induction to evolve from the "descent" form used by Euclid to the "ascent" form present in the Peano axioms.

Exercises

One of the oldest problems about numbers that is still not completely understood is the problem of Egyptian fractions. In ancient Egypt, fractions were always expressed as sums of reciprocals, for example

$$\frac{2}{3} = \frac{1}{2} + \frac{1}{6}.$$

It is not obvious that any fraction between 0 and 1 can be expressed in this form and, indeed, the Egyptians probably did not know this for a fact. But it turns out that the following naive method always works: given a fraction a/b between 0 and 1, subtract from a/b the largest reciprocal $1/c$ that is less than a/b, then repeat the process with the fraction $a/b - 1/c = a'/bc$. This method, which was used by Fibonacci (1202), always works because $a' < a$. So the process terminates in a finite number of steps (and it expresses m/n as a sum of *distinct* reciprocals).

2.1.1 Use the Fibonacci method to express $3/7$ as a sum of distinct reciprocals.
2.1.2 If $0 < a/b < 1$ and $1/c$ is the largest reciprocal less than a/b, show that $ac > b > a(c - 1)$.
2.1.3 Deduce from Exercise 2.1.2 that $0 < a' < a$, where

$$\frac{a}{b} - \frac{1}{c} = \frac{ac - b}{bc} = \frac{a'}{bc}.$$

Thus it is always possible to express a fraction as a sum of reciprocals, as the Egyptians wanted. However, not much is known about the number of reciprocals required, or how large their denominators may become.

The book (Fibonacci 1202) is also the source of the sequence of *Fibonacci numbers*,

$$0, 1, 1, 2, 3, 5, 8, 13, 21, 34, 55, 89, 144, 233, 377, 610, 987, \ldots,$$

in which each number (after the first two) is the sum of the preceding two.

2.1.4 Show that the Euclidean algorithm, applied to the pair of Fibonacci numbers $\langle 13, 8 \rangle$, terminates at the pair $\langle 1, 1 \rangle$.
2.1.5 Explain why the Euclidean algorithm, applied to any pair of consecutive Fibonacci numbers, terminates at the pair $\langle 1, 1 \rangle$.
2.1.6 Deduce that the greatest common divisor of any pair of consecutive Fibonacci numbers is 1.

Another descent process that necessarily terminates is that of *subtracting from n the largest power of 2 less than or equal to n*.

2.1.7 Use subtraction of the largest power of 2 to prove that each positive integer n can be expressed uniquely as a sum of distinct powers of 2.
2.1.8 What does Question 2.1.7 have to do with binary notation?

2.2 Induction and Arithmetic

Although induction has been present in mathematics at least since the time of Euclid, the realization that it underlies even "trivial" properties of addition and multiplication is quite recent. Even more remarkable, this discovery was published in a book intended for high school students, the *Lehrbuch der Arithmetik* (Textbook of Arithmetic) of Hermann Grassmann, in 1861. Grassmann's "new math" was far ahead of its time and it went unnoticed, even by mathematicians, until it was rediscovered by Dedekind (1888) and Peano (1889). It is still surprising to see how tightly induction is bound up with the properties of addition and multiplication that seem obvious from a visual point of view, such as $a + b = b + a$ and $ab = ba$.

We follow Grassmann by using the *natural numbers* $0, 1, 2, 3, \ldots$, rather than the positive integers $1, 2, 3, \ldots$, as it is more convenient to start with 0 when defining addition and multiplication. We also use $S(n)$, rather than $n + 1$, to denote the successor of n, in order to avoid any suspicion of circularity in defining addition. Thus we initially denote the natural numbers by $0, S(0), S S(0), S S S(0), \ldots$, and the successor function S is the only function we know.

A *proof by induction* that property P for all natural numbers n proceeds by proving the *base step*, that P holds for 0, and the *induction step*, that if P holds for $n = k$ then P holds for $n = S(k)$.

2.2.1   Addition

On this slender foundation we now build the addition function + by the following *inductive definition*:

$$m + 0 = m,$$

$$m + S(n) = S(m + n).$$

The first line defines $m + 0$ for all natural numbers m, while the second defines $m + S(n)$ for all m, assuming that $m + n$ has already been defined for all m. It follows, by induction, that $m + n$ is defined for natural numbers m and n. The first thing to notice about this definition of addition is that it implies $n + 1 = S(n)$, as it should, because 1 is defined to be $S(0)$ and so

$$n + 1 = n + S(0)$$

$$= S(n + 0) \qquad\qquad\qquad\qquad \text{by definition of addition,}$$

$$= S(n) \qquad\qquad \text{because } n + 0 = n \text{ by definition of addition.}$$

Next, $S(n) = 1 + n$, by the following induction on n. For $n = 0$ we have

$$S(0) = 1 = 1 + 0, \qquad\qquad\qquad\qquad \text{by definition of addition.}$$

And assuming that $S(n) = 1 + n$ holds for $n = k$ we have

$$1 + S(k) = S(1 + k) \qquad\qquad \text{by definition of addition,}$$
$$= SS(k) \qquad\qquad \text{by induction hypothesis,}$$
$$= S(k) + 1 \qquad\qquad \text{because } S(n) = n + 1.$$

So $S(n) = n + 1 = 1 + n$ for all natural numbers n, by induction.

Now we can use the definition to prove the algebraic properties of addition. A good illustration is the *associative law*,

$$l + (m + n) = (l + m) + n \qquad\qquad \text{for all natural numbers } l, m, n.$$

We prove this for all l and m by induction on n. First, the base step $n = 0$. In this case, the left side is $l+(m+0)$, which equals $l+m$ by definition of $m+0$. And the right side $(l + m) + 0$ also equals $l + m$, for the same reason. Thus $l + (m + n) = (l + m) + n$ holds for all l and m when $n = 0$.

Now for the induction step: we suppose that $l + (m + n) = (l + m) + n$ holds for $n = k$, and prove that it also holds for $n = S(k)$. Well,

$$l + (m + S(k)) = l + S(m + k) \qquad\qquad \text{by definition of addition,}$$
$$= S(l + (m + k)) \qquad\qquad \text{by definition of addition,}$$
$$= S((l + m) + k) \qquad\qquad \text{by induction hypothesis,}$$
$$= (l + m) + S(k) \qquad\qquad \text{by definition of addition.}$$

This completes the induction.

With the help of the associative law we can prove other algebraic properties of addition, such as the *commutative law*, $m + n = n + m$. The steps are outlined in the exercises below.

2.2.2 Multiplication

Now that we have the addition function, we can define multiplication, written $m \cdot n$ or simply mn, by the following induction:

$$m \cdot 0 = 0,$$
$$m \cdot S(n) = m \cdot n + m.$$

The usual algebraic properties of multiplication follow from this definition. For example, we can prove the *identity property*, that $1 \cdot m = m$ by induction on m as follows.

The base step $1 \cdot 0 = 0$ follows from the definition of multiplication. For the inductive step, we suppose that $1 \cdot m = m$ holds for $m = k$. Then

$$1 \cdot S(k) = 1 \cdot k + 1 \qquad \text{by definition of multiplication,}$$

$$= k + 1 \qquad \text{by induction hypothesis,}$$

$$= S(k) \qquad \text{by the proof above that } k + 1 = S(k).$$

This completes the induction, so $1 \cdot m = m$ for all natural numbers m.

2.2.3   The Law $ab = ba$ Revisited

Other familiar algebraic properties of multiplication follow by induction, as outlined in the exercises below. The main difficulty in these proofs is the absence of algebraic assumptions in the definitions of addition and multiplication. So all algebraic properties must be proved from scratch, and it is not obvious which ones to prove first. It turns out, for example, that one needs to prove associativity before proving commutativity.

This may be surprising, because we saw in Sect. 1.1 that $ab = ba$ is *obvious* when we view numbers as lengths and take the product of lengths a and b to be their "rectangle." But in Sect. 1.2 we saw that the concept of "length" is not as simple as it looks. In particular, it is not clear how to represent irrational lengths by numbers. If we wish to put the number concept on a sound foundation, we should presumably begin with the natural numbers, where Grassmann's approach allows us to *prove* $ab = ba$ without appeal to geometric intuition. Hopefully, we can then extend the number concept far enough to capture the concept of irrational length, while at the same time extending the proof that $ab = ba$ without appeal to geometric interpretations of a, b, and ab.

In the remainder of this chapter we explain how this program may be carried out. The number concept is extended in two stages; from natural numbers to *rational* numbers, and from rational numbers to *real* numbers. The first stage is fairly straightforward, and purely algebraic. The second stage is where we make the leap from the discrete to continuous, avoiding the use of geometric concepts by introducing a concept that is more universal: the *set* concept.

Exercises

Our first goal is to prove the commutativity of addition, $m + n = n + m$, by induction on n. This depends on some of the properties of addition already proved above.

2.2.1   Show that the base step, for $n = 0$, depends on proving that $0 + m = m$, and prove this by induction on m.

2.2.2 The induction step is that if $m + k = k + m$ then $m + S(k) = S(k) + m$. Prove this implication, using the results $m + 1 = 1 + m$, $S(k) = k + 1$ and associativity proved above.

Next, in order to deal with combinations of addition and multiplication, it will be useful to have the *left distributive law*, $l(m + n) = lm + ln$, and the *right distributive law*, $(l + m)n = ln + mn$.

2.2.3 Prove the left distributive law by induction on n, using the definition of multiplication and the associativity of addition.

2.2.4 Prove the *associative law of multiplication*, $l(mn) = (lm)n$, by induction on n. For the induction step use the definition of multiplication and the left distributive law.

2.2.5 Prove the right distributive law by induction on n, using the definition of multiplication and the associative and commutative laws for addition.

Finally we are ready to prove the *commutative law of multiplication*, $mn = nm$ by induction on n. (One wonders why this result is so hard to reach by induction, when it seemed so easy in Sect. 1.1. Apparently, $mn = nm$ can be true for reasons quite different from those that first come to mind. See also Sect. 2.9.)

2.2.6 Show that the base step, $n = 0$, follows from $0 \cdot m = 0$, and prove the latter by induction on m.

2.2.7 Show that the induction step, $mk = km$ implies $m \cdot S(k) = S(k) \cdot m$, follows from the definition of multiplication, identity property, and right distributive law.

2.3 From Rational to Real Numbers

The arithmetic of natural numbers in the previous section is easily extended to the arithmetic of non-negative rational numbers m/n, where m and n are any natural numbers with $n \neq 0$. Admittedly, the definition of sum for rational numbers

$$\frac{a}{b} + \frac{c}{d} = \frac{ad + bc}{bd}$$

is quite sophisticated, and it causes a lot of grief in elementary school. But once this concept is mastered, it is not hard to see that the commutative, associative laws, and so on, extend from the natural numbers to the rational numbers m/n. Moreover, the rational numbers have a convenient property that the natural numbers lack: each rational number $r \neq 0$ has a *multiplicative inverse* r^{-1} such that $rr^{-1} = 1$. Namely, if $r = m/n$ then $r^{-1} = n/m$.

Thus there is not much difference between the arithmetic of natural numbers and that of the rational numbers. To extend arithmetic to *irrational* numbers we need to understand where the irrational numbers lie relative to the set of rational numbers, and we really have to start thinking of the rational numbers as a set, which we denote by $\mathbb{Q}$. (The symbol $\mathbb{Q}$ apparently stands for "quotients.")

In Sects. 1.2 and 1.3 we observed the difficulties raised by the existence of irrational numbers, such as $\sqrt{2}$, for geometry and arithmetic. On the one hand, we want enough numbers to fill the line. On the other hand, we want to be able to add and multiply the points on the line in a way that is consistent with addition and multiplication of rational numbers.

In Sect. 1.3 we floated the idea of using infinite decimals to represent irrational numbers, but immediately cast doubts on its practicality, due to the difficulty of adding and multiplying infinite decimals. This makes it hard to tell whether their arithmetic is even compatible with the arithmetic of rational numbers. For example, how would you like to verify that

$$\frac{1}{6} \times \frac{1}{7} = \frac{1}{42},$$

using the infinite decimals

$$\frac{1}{6} = 0.166666666666\ldots$$

$$\frac{1}{7} = 0.142857142857\ldots \quad ?$$

Nevertheless, infinite decimals *do* solve the problem of representing all points of the line, so it is worth exploring them a little further. We will see that infinite decimals can serve as a stepping-stone to a concept of *real numbers* that is not only faithful to the image of points on a line, but also compatible with the arithmetic of rational numbers. Moreover, the new real number concept comfortably includes numbers such as $\sqrt{2}$, and allows us to prove results as $\sqrt{2} \times \sqrt{3} = \sqrt{6}$.

First, let us revisit the infinite decimal for $\sqrt{2}$, as it will help to explain why there are enough infinite decimals to fill a line. The infinite decimal

$$\sqrt{2} = 1.414213\ldots$$

is what we call the *least upper bound* (lub) of the following set of finite decimal fractions:

$$1$$

$$1.4$$

$$1.41$$

$$1.414$$

$$1.4142$$

$$1.41421$$

$$1.414213$$

$$\vdots$$

It is *an* upper bound because it is greater than each of them, and it is *least* because any number less than $1.414213\ldots$ must be less in some decimal place, and hence less than some member of the set. Thus any number less than 1.414213 . . . is not an upper bound at all.

The number 1.414213 . . . is also the *greatest lower bound* (glb) of the set

$$2$$

$$1.5$$

$$1.42$$

$$1.415$$

$$1.4143$$

$$1.41422$$

$$1.414214$$

$$\vdots$$

—by a similar argument. Thus the irrational number $\sqrt{2}$ fills a "point-sized hole," or *gap*, between the two sets of finite decimals

$$1, \quad 1.4, \quad 1.41, \quad 1.414, \quad 1.4142, \quad 1.41421, \quad 1.414213, \quad \ldots$$

and

$$\ldots, \quad 1.414214, \quad 1.41422, \quad 1.4143, \quad 1.415, \quad 1.42, \quad 1.5, \quad 2.$$

There are of course many holes in the set of finite decimal fractions (a simpler example is 1/3), but each of them is a "point-sized hole" for the same reason that $\sqrt{2}$ is: we can approach it arbitrarily closely, from below or above, by finite decimal fractions. Thus the infinite decimals *complete* the number line by filling all the gaps.

Looking back over this explanation, one sees that there is nothing special about the set of finite decimal fractions. We could use any fractions that are plentiful enough to approach each point arbitrarily closely. For example, the *binary fractions* $m/2^n$ (for integers m, n) suffice. At the other extreme, one could use all the rational numbers m/n. This gives the advantage of a simple approach to addition and multiplication, as we will see in the next section. So let us see what gaps look like in the set of rational numbers. It is convenient to consider just the set $\mathbb{Q}^+$ of non-negative rationals for now.

A gap occurs wherever the set $\mathbb{Q}^+$ breaks into two sets, L and U, with the following properties.

1. Each member of L is less than every member of U.
2. L has no greatest member and U has no least member.

Under these circumstances, the gap between L and U represents a single irrational number, equal to lub L and glb U. An example is where

$$U = \{r \in \mathbb{Q}^+ : r^2 > 2\},$$

in which case the gap represents the number $\sqrt{2} = \text{glb } U$.

A separation of $\mathbb{Q}^+$ into two sets L and U with the above properties is called a *Dedekind cut* after Richard Dedekind, who first thought of representing irrational numbers in this way. His cuts occur exactly where the gaps are. You could even say that the cut *is* the gap, so we are filling the gaps simply by recognizing the gaps as new mathematical objects!

This is not a joke, but actually a deep idea. Since L and U completely determine an irrational number, we can take this pair of sets to *be* an irrational number. We create a new mathematical object by comprehending a collection of existing mathematical objects as a *set*. Dedekind was the first to notice the power of set comprehension, and in doing so he launched the program of *arithmetization*—building all of mathematics on the foundation of natural numbers and sets.

Since either L or U completely determines the cut, it suffices to represent the corresponding irrational number simply by L. This is convenient because then *the ordering of real numbers corresponds to set containment*. That is,

$$\text{lub } L \leq \text{lub } L' \Leftrightarrow L \subseteq L'.$$

Pursuing this idea a little further, we note that each rational number s is the lub of a set L of rational numbers, namely

$$L = \{r \in \mathbb{Q}^+ : r < s\},$$

and L has the following properties of the lower part of a Dedekind cut:

$1'$. L is a bounded set of positive rationals with no greatest member.
$2'$. L is "closed downward," that is, if $p \in L$ and $0 < q < p$, then $q \in L$.

Thus, if we define a *lower Dedekind cut* to be any set L with properties $1'$ and $2'$, then every positive real number, rational or irrational, can be represented by a lower Dedekind cut. We thereby obtain a uniform representation of positive real numbers as certain sets of rational numbers, and set containment gives the usual ordering of numbers. In the next section we will see how lower Dedekind cuts may also be used to define sums, products, and square roots of positive numbers.

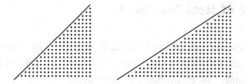

Fig. 2.1 The lower Dedekind cuts for 1 and 2/3

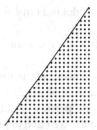

Fig. 2.2 The lower Dedekind cut for $\sqrt{2}$

2.3.1 Visualizing Dedekind Cuts

To visualize Dedekind cuts we first "spread out" the rationals as we did in Sect. 1.3. That is, we view n/m as the integer point $\langle m, n \rangle$ in the plane (the point at *slope n/m* from $\langle 0, 0 \rangle$). Then the lower Dedekind cut corresponding to a real number r is seen as the set of integer points below the line through $\langle 0, 0 \rangle$ with slope r. Figure 2.1 shows the lower Dedekind cuts for 1 and 2/3 in this fashion.

Figure 2.2 shows the cut for $\sqrt{2}$. Since points are shown as small disks, "points" may appear to fall on the line though they are actually below it. [It is a good exercise to check this for some points, such as $\langle 7, 5 \rangle$.]

Exercises

As we remarked earlier, an advantage of decimals is that they instantly tell us which is the larger of two numbers; we have only to look at the first decimal place (from the left) where the two decimals differ. This advantage extends to describing the least upper bound of a set of numbers given by infinite decimals.

2.3.1 Suppose that S is a set of real numbers between 0 and 1, and l is the least upper bound of S. How is the first decimal place of l determined by the first decimal places of the members of S?

2.3.2 How is the second decimal place of l determined?

2.3.3 Using Questions 2.3.1 and 2.3.2, give a description of the decimal expansion of l.

2.3.4 Does a similar idea work to find the glb of S?

2.4   Arithmetic of Real Numbers

Thanks to the representation of positive real numbers by lower Dedekind cuts, all properties of real numbers reduce to properties of rational numbers, which we already understand. For example, we can explain the addition of positive real numbers in terms of addition of rational numbers by means of the following definition.

Definition. If L and L' are lower Dedekind cuts, then $L + L'$ is defined by

$$L + L' = \{r + r' : r \in L \text{ and } r' \in L'\}.$$

In other words, the sum of two lower cuts is the set of sums of their respective members. Notice that we immediately have $L + L' = L' + L$, because $r + r' = r' + r$ for rational numbers. Other algebraic properties of cuts are similarly "inherited" from those of rational numbers. Admittedly, we have to prove that the sets obtained in this way are themselves lower Dedekind cuts. This turns out to be fairly straightforward and, as promised, it depends only on properties of rational numbers.

Sum theorem for Dedekind cuts. *If L and L' are lower Dedekind cuts, then so is $L + L'$, and it agrees with the ordinary sum on the rational numbers.*

Proof. The sum: $L + L' = \{r + r' : r \in L \text{ and } r' \in L'\}$ is certainly a set of rational numbers, bounded above by the sum of bounds on L and L'. Now suppose that $r + r' \in L + L'$ and that t is a rational number less than $r + r'$. We have to show that t is also in $L + L'$, in other words that $t = s + s'$, for some $s \in L$ and $s' \in L'$. One way to do this is to divide t into two pieces in the ratio of r to r': that is, let

$$s = \frac{tr}{r + r'}, \quad s' = \frac{tr'}{r + r'}.$$

Then s is rational and $s < r$, so $s \in L$; similarly $s' \in L'$; and clearly $s + s' = t$.

In the special case where L and L' represent rational numbers, l and l' say,

$$L = \{r \in \mathbb{Q}^+ : r < l\}, \quad L' = \{r' \in \mathbb{Q}^+ : r' < l'\}.$$

Then, for any $r \in L$ and $r' \in L'$ we have $r + r' < l + l'$. Conversely, for any $t < l + l'$ we have $t = r + r'$ with $r < l$ and $r' < l'$; namely, let $r = \frac{l}{l+l'}t$ and $r' = \frac{l'}{l+l'}t$.   □

We similarly would say that the product of two lower cuts is the set of products of their members.

Definition. If L and L' are lower Dedekind cuts, then LL' is defined by

$$LL' = \{rr' : r \in L \text{ and } r' \in L'\}.$$

Product of Dedekind cuts. *If L and L' are lower Dedekind cuts, then so is LL', and it agrees with the ordinary product on the rational numbers.*

Proof. The product $LL' = \{rr' : r \in L \text{ and } r' \in L'\}$ is a set of rationals, bounded above by the product of the bounds on L and L'. Now suppose that $rr' \in LL'$ and that t is a rational less than rr'. To show that t is also in LL' we have to find $s \in L$ and $s' \in L'$ such that $t = ss'$. Since $\frac{t}{rr'} < 1$, there is a rational q such that

$$\frac{t}{rr'} < q < 1,$$

so we can take

$$s = rq, \quad \text{which is less than } r \text{ because } q < 1,$$

$$s' = \frac{t}{rq}, \quad \text{which is less than } r' \text{ because } \frac{t}{rr'} < q,$$

and this obviously gives $t = ss'$.

In the special case where L and L' represent rational numbers, l and l' say,

$$L = \{r \in \mathbb{Q}^+ : r < l\}, \quad L' = \{r \in \mathbb{Q}^+ : r < l'\}.$$

Then, for any $r \in L$ and $r' \in L'$ we have $rr' < ll'$. Conversely, for any $t < ll'$ we have $t = rr'$ with $r < l$ and $r' < l'$. Namely, choose a rational q with

$$\frac{t}{ll'} < q < 1,$$

and let

$$r = lq, \quad \text{which is less than } l \text{ because } q < 1,$$

$$r' = \frac{t}{lq}, \quad \text{which is less than } l' \text{ because } \frac{t}{ll'} < q.$$

This gives $t = rr'$ as required. $\qquad\qquad\square$

2.4.1 The Square Root of 2

We have already seen one valid way to describe $\sqrt{2}$: by an infinite decimal. In Sect. 2.7 we will see another way, by a continued fraction. But neither fits easily into an arithmetic theory of real numbers, because it is hard to describe multiplication of infinite decimals (and even harder for continued fractions). Dedekind cuts, on the other hand, are easily multiplied, and this leads to a relatively easy treatment of square roots. We show how in the case of $\sqrt{2}$.

Existence of $\sqrt{2}$. *The lower Dedekind cut* $K = \{s \in \mathbb{Q}^+ : s^2 < 2\}$ *represents* $\sqrt{2}$, *because*

$$K^2 = \{r \in \mathbb{Q}^+ : r < 2\},$$

which is the lower Dedekind cut representing 2.

Proof. By the definition of product of lower Dedekind cuts,

$$K^2 = \{ss' : s, s' \in K\} = \{ss' : s^2, s'^2 < 2\}.$$

We know, from the product theorem above, that K^2 is a lower Dedekind cut, so it suffices to prove that $\mathrm{lub}(K^2) = 2$.

Well, if $s^2, s'^2 < 2$, then $ss' \leq$ one of $s^2, s'^2 < 2$, so $\mathrm{lub}(K^2)$ is *at most* 2. To show that $\mathrm{lub}(K^2)$ is *at least* 2 it suffices to show the following: for each rational $r < 2$ there is a rational s with $r < s^2 < 2$ (because in that case $s \in K$ and $s^2 \in K^2$).

Such an s^2 can always be found[1] by choosing $s \in K$ sufficiently close to $\mathrm{lub}(K)$, which is possible because there are rational numbers arbitrarily close to $\mathrm{lub}(K)$.   □

2.4.2   The Equation $\sqrt{2}\,\sqrt{3} = \sqrt{6}$

Dedekind (1872) wrote (in the 1901 English translation, p. 22)

> Just as addition is defined, so can the other operations of the so-called elementary arithmetic be defined ... differences, products, quotients, powers, roots, logarithms, and in this way we arrive at proofs of theorems (as, e.g., $\sqrt{2} \cdot \sqrt{3} = \sqrt{6}$, which to the best of my knowledge have never been established before).

However, Dedekind did not go beyond defining the sum of cuts, so the proof that $\sqrt{2}\,\sqrt{3} = \sqrt{6}$ is not in his book either. Now that we have defined the product of cuts, and found the cut for $\sqrt{2}$, we are very close to such a proof. It remains to define $\sqrt{3}$, by the cut

$$K' = \{s \in \mathbb{Q}^+ : s^2 < 3\},$$

---

[1] A specific way to do this is to choose rational s and t that are close together on either side of $\mathrm{lub}(K)$. For example, find $s \in K$, $t \notin K$ with $t < 2$ and $t - s < \frac{2-r}{4}$. Then we have

$$t^2 - s^2 = (t + s)(t - s) < 4 \cdot \frac{2 - r}{4} = 2 - r,$$

and hence $r < s^2 < 2 < t^2$.

and to prove that

$$K'^2 = L' = \{r \in \mathbb{Q}^+ : r < 3\}$$

(the cut representing 3). This is very similar to the proof $K^2 = L$ above, and we leave it as an exercise.

Then $\sqrt{2}\sqrt{3}$ is represented by the cut KK', with the square

$$(KK')^2 = K^2 K'^2 = LL',$$

which represents $2 \cdot 3 = 6$. So KK', which represents $\sqrt{2}\sqrt{3}$, also represents $\sqrt{6}$.

Exercises

The following two exercises verify that 3 is the square of the lower Dedekind cut $K' = \{s \in \mathbb{Q}^+ : s^2 < 3\}$, as assumed in the proof above.

2.4.1 Show that, for each rational $r < 3$, there is a rational s with $r < s^2 < 3$. (*Hint:* Consider $s \in K'$ and $t \notin K'$ with $t < 2$ and $t - s < \frac{3-r}{4}$.)

2.4.2 Deduce that $K'^2 = \{r \in \mathbb{Q}^+ : r < 3\}$, so K' represents $\sqrt{3}$.

By generalizing this argument we can show the existence of roots of all real numbers, and other algebraic properties.

2.4.3 Explain why each positive real number has a square root.
2.4.4 Explain why each positive real number has a cube root.
2.4.5 Show that properties such as $ab = ba$ (used in the proof that $\sqrt{2}\sqrt{3} = \sqrt{6}$) are inherited from the rational numbers.

2.5 Order and Algebraic Properties

Now we come back to the question asked in Sect. 1.3: what are points, and how do they fill the line? For simplicity we will consider just the positive real numbers that we have defined so far. These are the "points" (members) of a set $\mathbb{R}^+$ we will call the *positive number line*. (It is no secret that we are going to introduce negative numbers shortly, so as to obtain the full number line $\mathbb{R}$.)

The properties of $\mathbb{R}^+$ that make it a "line" are called *order* properties and they can be expressed in terms of the $\leq$ relation (which corresponds to the containment relation $\subseteq$ between lower Dedekind cuts). The following properties of $\leq$ define what we call a *linear order*. For any $a, b, c \in \mathbb{R}$:

$$a \leq a,$$

$$\text{either } a \leq b \text{ or } b \leq a,$$

$$\text{if } a \leq b \text{ and } b \leq c \text{ then } a \leq c.$$

All of these properties are obvious for positive numbers because $\leq$ means $\subseteq$. However, a linearly ordered set of points is not necessarily what we would call a "line."

The natural numbers are linearly ordered, but they are *isolated* in the sense that for each natural number n there is an empty space before the next natural number $n + 1$, so the natural numbers come nowhere close to filling the line. The rational numbers come closer, because they lie *densely* on the line. That is, between any two rational numbers there is another. But, as we know, even this dense set has *gaps*. In particular, there is a gap in the rational numbers at the position $\sqrt{2}$. What this means, precisely, is that *the set* $\{r \in \mathbb{Q}^+ : r^2 < 2\}$ *has no least upper bound in* $\mathbb{Q}^+$, because the rationals r with $r^2 \geq 2$ have no least member.

$\mathbb{R}^+$ qualifies as a "line" because it is dense *and* has no gaps, because *any bounded set of real numbers has a least upper bound*.

This property follows from the fact that real numbers are (or are represented by) certain subsets of $\mathbb{Q}^+$, namely, lower Dedekind cuts. A bounded set of real numbers x is therefore a set of lower Dedekind cuts L_x that all lie below some bound r. But then the *union L* of all the cuts L_x is itself a lower Dedekind cut, and it is obviously the *least* cut that contains (which means $\geq$, remember) all the cuts L_x. Thus the union L of the cuts L_x is their least upper bound.[2]

The least upper bound property has important consequences in analysis, such as the intermediate value theorem and the integrability of continuous functions. We will study these results in Chap. 4. They depend on the fact that $\mathbb{R}$ has no gaps, which we call the *completeness* of $\mathbb{R}$.

The final definitive property of the order of $\mathbb{R}$ is called the *Archimedean* property. It says that, if a and b are numbers with $0 < a < b$, then $na > b$ for some natural number n. The property holds because

in the lower Dedekind cut for a there is a rational $p/q < a$,

in the upper Dedekind cut for b there is a rational $r/s > b$,

and we can certainly find a natural number n such that

$$np/q > r/s.$$

For example, $n = rq$ will do.

The Archimedean property implies that *there are no infinitesimal numbers*; that is, numbers $a > 0$ with $a < 1/n$ for all natural numbers n. If a is infinitesimal, then $na < 1$ for all natural numbers n, which contradicts the Archimedean property with $b = 1$.

To summarize: *the order of the positive real numbers is linear, dense, Archimedean, and complete.*

---

[2]In particular, when the bounded set consists of the cuts L_r where $r^2 < 2$, its least upper bound L is the cut representing $\sqrt{2}$.

2.5.1 Algebraic Properties of $\mathbb{R}$

Until now we have worked only with non-negative numbers, so as to take the shortest route to Dedekind cuts and their multiplication. Looking back at the route taken, it is easy to see how negative numbers can be carried along as well.

We began with the natural numbers, $0, 1, 2, 3, \ldots$, showing that they have the following algebraic properties (all provable by induction):

$$a + b = b + a \quad ab = ba \qquad \text{(commutative laws)}$$

$$a + (b + c) = (a + b) + c \quad a(bc) = (ab)c \qquad \text{(associative laws)}$$

$$a + 0 = a \quad a \cdot 1 = a \qquad \text{(identity laws)}$$

$$a(b + c) = ab + ac \qquad \text{(distributive law)}$$

When we adjoin the negative integer $-m$ for each positive m the above laws are preserved if we let $(-m)n = -mn$, and we gain the *additive inverse law*:

$$a + (-a) = 0.$$

These eight algebraic laws define what is called a *ring* (or more precisely, a commutative ring with unit). We call the natural numbers and their negatives the *integers*, and denote the ring of integers by $\mathbb{Z}$. The symbol comes from the German word "Zahlen" for "numbers."

The quotients m/n of integers m, n with $n \neq 0$ form the set $\mathbb{Q}$ of (positive and negative) *rational numbers*, and they satisfy all the laws of a ring, plus the *multiplicative inverse law*:

$$a \cdot a^{-1} = 1 \quad \text{for} \quad a \neq 0,$$

where $a^{-1} = n/m$ if $a = m/n$. A structure satisfying these nine laws is called a *field*.

Our final objective is to introduce negative real numbers so that the resulting set $\mathbb{R}$ of real numbers is a field. If we continue to define positive real numbers r as subsets of the set $\mathbb{Q}^+$ of positive rationals, we need to adjoin 0 and a negative real $-a$ for each positive a. Then the rule $(-a)b = -ab$ ensures that all the field laws continue to hold.

Now let us observe how the order of $\mathbb{R}$ interacts with sums and products. Clearly, we have

$$0 < 1,$$

$$\text{if } a < b \text{ then } a + c < b + c,$$

$$\text{if } a < b \text{ and } c > 0 \text{ then } ac < bc.$$

Given that $\leq$ is a linear order, as defined above, a field with the latter three properties is called an *ordered field*. Since the order of real numbers is complete we have: $\mathbb{R}$ is a *complete ordered field*.

It is not necessary to add that $\mathbb{R}$ is Archimedean, because this property actually follows from completeness in an ordered field.

Archimedean property of a complete ordered field. *If F is a complete ordered field and $a, b \in F$ with $0 < a < b$, then $na > b$ for some natural number n.*

Proof. Suppose, on the contrary, that $na \leq b$ for each natural number n. Then the set $\{a, 2a, 3a, \ldots\}$ is bounded and hence has a least upper bound c, by completeness.

Since $c - a < c$ it follows that $c - a < na$ for some natural number n. But then $c < (n + 1)a$ contrary to the definition of c. $\square$

Exercises

When one first meets infinite decimals, it seems hard to believe that $0.9999\ldots = 1$, because it seems that there should be an "infinitesimal" difference between 1 and $0.9999\ldots$.

2.5.1 Show, on the contrary, how this is a good illustration of the Archimedean property of real numbers.

We saw an example of a *non*-Archimedean linearly ordered set in the exercises to Sect. 1.5, namely the set $\mathcal{R}$ of rational functions with real coefficients, in which the constant functions represent the real numbers and the function $\iota(x) = 1/x$ is an infinitesimal.

2.5.2 Show that $\mathcal{R}$ is not complete, because the set of infinitesimal functions is bounded but has no least upper bound.

$\mathbb{R}$ is in fact the *only* complete ordered field ("up to isomorphism"), as the following exercises show. Suppose F is such a field.

2.5.3 Deduce from the properties $0 < 1$ and $a + c < b + c$ when $a < b$ that F contains elements

$$\cdots < -2 < -1 < 0 < 1 < 2 < \cdots,$$

and hence is a copy of the integers.

2.5.4 Deduce, by forming quotients, that F contains a copy of the rationals.

2.5.5 For each $x \in F$, consider the lower Dedekind cut L_x consisting of the rationals in F that are $< x$. Deduce from completeness that lub $L_x = x$, so the elements of F are in bijective correspondence with the real numbers.

2.6 Other Completeness Properties

Absence of gaps may be the most intuitive way to think of completeness, but often it is better to think of completeness as a guarantee that certain infinite processes lead to a result. Forming an infinite decimal is one such example. We now discuss two others, which often arise in analysis.

The first concerns closed intervals, the sets of the form

$$[a, b] = \{x \in \mathbb{R} : a \leq x \leq b\},$$

and the second (which follows from the first) concerns *convergence of sequences*. We will show only that these two results follow from the lub (and glb) property; however, it can also be shown that they imply it.

Nested Interval Property. *If $I_1 \supseteq I_2 \supseteq I_3 \supseteq \cdots$ are closed intervals with lengths that become arbitrarily small, then $I_1, I_2, I_3, \ldots$ have a single common point.*

Proof. Let $I_1 = [a_1, b_1], I_2 = [a_2, b_2], \ldots,$ so we have

$$a_1 \leq a_2 \leq a_3 \leq \cdots \leq b_3 \leq b_2 \leq b_1.$$

It follows, by completeness, that $\text{lub}\{a_1, a_2, a_3, \ldots\}$ and $\text{glb}\{b_1, b_2, b_3, \ldots\}$ both exist. We therefore have

$$a_1 \leq a_2 \leq a_3 \leq \cdots \leq \text{lub}\{a_1, a_2, a_3, \ldots\}$$

$$\leq \text{glb}\{b_1, b_2, b_3, \ldots\} \leq \cdots \leq b_3 \leq b_2 \leq b_1.$$

Thus any x in the interval from $\text{lub}\{a_1, a_2, a_3, \ldots\}$ to $\text{glb}\{b_1, b_2, b_3, \ldots\}$ is common to all of $I_1, I_2, \ldots$.

If the length of the intervals $I_1, I_2, \ldots$ becomes arbitrarily small, then

$$x = \text{lub}\{a_1, a_2, a_3, \ldots\} = \text{glb}\{b_1, b_2, b_3, \ldots\}$$

is the *only* common point.   □

We now use nested intervals to study limit points of sequences.

Definition. A sequence of numbers $c_1, c_2, c_3, \ldots$ *converges* if it has a *limit* c, that is, if c_n becomes arbitrarily close to c as n increases. More precisely, c is the limit of the sequence $c_1, c_2, c_3, \ldots$ if, for each number $\varepsilon > 0$, there is a natural number N such that

$$n > N \implies |c - c_n| < \varepsilon.$$

We would like to be able to tell whether a sequence converges without knowing in advance what its limit is. The *Cauchy convergence criterion* makes this possible.

Cauchy convergence criterion. *Sequence $c_1, c_2, c_3, \ldots$ converges if, for each $\varepsilon > 0$, there is an N such that*

$$m, n > N \implies |c_m - c_n| < \varepsilon.$$

Proof. If the sequence $c_1, c_2, c_3, \ldots$ satisfies the Cauchy convergence criterion there is a sequence of natural numbers $N_1 < N_2 < N_3 < \cdots$ such that

$$m, n > N_1 \;\Rightarrow\; |c_m - c_n| < 1/2,$$
$$m, n > N_2 \;\Rightarrow\; |c_m - c_n| < 1/4,$$
$$m, n > N_3 \;\Rightarrow\; |c_m - c_n| < 1/6,$$

and so on. Now if $|c_m - c_n| < 1/2$ for all $m, n > N_1$ this means in particular that all c_n stay within distance $1/2$ of c_{N_1+1} for $n > N_1$, and hence within an interval of length 1. Similarly for $N_2, N_3, \ldots$; so we get nested closed intervals

$$I_1, \text{ of length 1, with } c_n \in I_1 \text{ for all } n > N_1,$$

$$\supseteq I_2, \text{ of length } 1/2, \text{ with } c_n \in I_2 \text{ for all } n > N_2,$$

$$\supseteq I_3, \text{ of length } 1/3, \text{ with } c_n \in I_3 \text{ for all } n > N_3,$$

$$\vdots$$

the length of which becomes arbitrarily small. By the nested interval property, there is a single point c common to these intervals, and c is clearly the limit of the sequence $c_1, c_2, c_3, \ldots$.   □

We see from this proof that the Cauchy convergence criterion guarantees a limit because of the nested interval property, and hence ultimately because of the completeness of $\mathbb{R}$. Conversely, as mentioned above, if each sequence satisfying the Cauchy criterion has a limit then $\mathbb{R}$ is complete. Indeed the completeness of $\mathbb{R}$ is often expressed this way: every *Cauchy sequence* has a limit, where a Cauchy sequence is one satisfying the Cauchy convergence criterion. This may seem more longwinded than, say, the least upper bound property, but it is important for several reasons.

One is that sequences of numbers are very common—we have already seen several examples—so we need to understand the concept of convergence. Moreover, numbers can very well be complex, and hence not ordered, so the concept of Dedekind cut may not apply. Another reason is that sequences of *functions* are also common, and we can use the Cauchy convergence criterion for functions, where the concept of Dedekind cut also does not generally apply.

Exercises

2.6.1 Define a sequence of nested *open* intervals with no common point.

The nested interval property and the Cauchy convergence criterion are both nicely illustrated by infinite decimals.

2.6.2 Interpret the infinite decimal for $\sqrt{2}$ as the common point of a sequence of nested intervals.

2.6.3 Show that the sequence $1, 1.4, 1.414, 1.4142, 1.41421, \ldots$, which defines the infinite decimal for $\sqrt{2}$, satisfies the Cauchy convergence criterion.

Consider nested sequences of intervals $I_1 \supseteq I_2 \supseteq I_3 \supseteq \cdots$ with lengths that converge to zero, where $I_n = [a_n, b_n]$ and a_n, b_n are rational.

2.6.4 Show that each such sequence corresponds to a Dedekind cut.

2.6.5 Show that two such sequences have the same common point if and only if they correspond to the same Dedekind cut.

2.6.6 Deduce that if real numbers are *defined* by such sequences, then we get real numbers with the same properties as those defined by Dedekind cuts.

2.6.7 Deduce in turn that real numbers can be defined by Cauchy sequences.

2.7 Continued Fractions

The Euclidean algorithm from Sect. 2.1, which operates on a pair $\langle a, b \rangle$ of positive integers, can also be viewed as a procedure for expressing each positive rational a/b as a *continued fraction*. Indeed, the continued fraction elegantly encodes the main steps in the algorithm.

Here is an example. We operate on the numbers 19 and 7, by first encoding them in the fraction 19/7.

$$\frac{19}{7} = 2 + \frac{5}{7} \qquad\qquad \text{subtracting 7 twice from 19,}$$

$$= 2 + \frac{1}{7/5} = 2 + \cfrac{1}{1 + \cfrac{2}{5}} \qquad\qquad \text{subtracting 5 once from 7,}$$

$$= 2 + \cfrac{1}{1 + \cfrac{1}{5/2}} = 2 + \cfrac{1}{1 + \cfrac{1}{2 + \cfrac{1}{2}}} \qquad\qquad \text{subtracting 2 twice from 5.}$$

The numbers $2, 1, 2, 2$ occurring in the continued fraction for 19/7 record the number of times the smaller number can be subtracted from the larger at each stage. In other words, they record the *quotient* when the larger number is *divided* by the smaller.

Given any number m/n, where m and n are positive integers, we can similarly show that

$$\frac{m}{n} = n_1 + \cfrac{1}{n_2 + \cfrac{1}{\ddots \cfrac{1}{n_{k-1} + \cfrac{1}{n_k}}}}$$

where $n_1, n_2, \ldots, n_k$ are positive integers. A fraction of this form is called a *finite continued fraction*. The fraction terminates because the Euclidean algorithm terminates.

To prove that $\sqrt{2}$ is irrational it therefore suffices to show that the Euclidean algorithm does *not* terminate on the pair of numbers $\sqrt{2}, 1$. Surprisingly, this is not hard to do. Here is what happens. The secret is to use the fact that $(\sqrt{2}+1)(\sqrt{2}-1) = 1$. As above, whenever we get a number less than 1 we rewrite it as 1/(number greater than 1), in order to continue the fraction.

$$\sqrt{2} = 1 + \sqrt{2} - 1 \qquad\qquad \text{subtracting 1 once from } \sqrt{2},$$

$$= 1 + \frac{1}{\sqrt{2} + 1} \qquad\qquad \text{because } (\sqrt{2} + 1)(\sqrt{2} - 1) = 1,$$

$$= 1 + \cfrac{1}{2 + \sqrt{2} - 1} \qquad\qquad \text{subtracting 1 twice from } \sqrt{2} + 1,$$

$$= 1 + \cfrac{1}{2 + \cfrac{1}{\sqrt{2} + 1}} \qquad\qquad \text{because } (\sqrt{2} + 1)(\sqrt{2} - 1) = 1.$$

At this point it is clear that the Euclidean algorithm will not terminate, because the denominator $\sqrt{2} + 1$ has occurred previously.

It follows that $\sqrt{2}$ is not equal to any ratio m/n of positive integers. So we have again proved that $\sqrt{2}$ is irrational—and this time without using proof by contradiction.

Moreover, we now have a clearer view of the irrational number $\sqrt{2}$. It can be described by a simple repetitive process, the Euclidean algorithm on $\sqrt{2}$ and 1. And this gives a simple repetitive *formula* for $\sqrt{2}$, namely its *infinite continued fraction*:

$$\sqrt{2} = 1 + \cfrac{1}{2 + \cfrac{1}{2 + \cfrac{1}{2 + \cfrac{1}{2 + \cfrac{1}{\ddots}}}}}$$

If the fraction x makes sense (which it does, as we will prove rigorously below) then it is much more transparent than the infinite decimal for $\sqrt{2}$, because we can survey its totality: a 1 followed by infinitely many 2s. In this sense, it is as transparent as an ultimately periodic decimal, such as the decimal $0.166666\ldots$ that represents $1/6$. For more on ultimate periodicity in continued fractions, see the exercises below.

Exercises

The simplest infinite continued fraction represents the famous number $\frac{1+\sqrt{5}}{2}$ known as the *golden ratio*. The golden ratio is the ratio of the sides of the *golden rectangle*, shown in Fig. 2.3. Its defining property is that the rectangle obtained by cutting off a square has the same shape as the original.

When a square is cut off as shown in Fig. 2.3, the width of the rectangle that remains is of course the greater side minus the lesser. So, by repeating the process of cutting off squares, we can implement the Euclidean algorithm.

2.7.1   Prove, from the defining property of the golden rectangle, that the golden ratio equals $\frac{1+\sqrt{5}}{2}$.

2.7.2   Prove that the Euclidean algorithm does not terminate on the pair $\langle \frac{1+\sqrt{5}}{2}, 1 \rangle$ by considering the golden rectangle.

2.7.3   Deduce that $\frac{1+\sqrt{5}}{2}$ is irrational.

2.7.4   What is the continued fraction for $\frac{1+\sqrt{5}}{2}$?

Periodic continued fractions, such as the one for $\frac{1+\sqrt{5}}{2}$, can be evaluated by showing that they satisfy quadratic equations. Here is another example. Let

$$x = 3 + \cfrac{1}{3 + \cfrac{1}{3 + \cfrac{1}{3 + \cfrac{1}{\ddots}}}}.$$

2.7.5   Show that x satisfies the equation $x^2 = 3x + 1$, and hence find x.

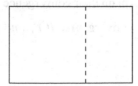

Fig. 2.3   The golden rectangle

2.8   Convergence of Continued Fractions

The nested interval property from Sect. 2.6 may be used to clarify the nature of
infinite continued fractions, which we introduced in the previous section without
investigating exactly what they mean. We define the infinite continued fraction

$$a_0 + \cfrac{1}{a_1 + \cfrac{1}{a_2 + \cfrac{1}{\ddots}}} \qquad \text{for any positive integers } a_0, a_1, a_2, \ldots,$$

to be the limit of the sequence of finite continued fractions

$$c_0 = a_0, \quad c_1 = a_0 + \frac{1}{a_1}, \quad c_2 = a_0 + \cfrac{1}{a_1 + \cfrac{1}{a_2}}, \quad \ldots,$$

which are known as the *convergents* of the infinite continued fraction. We are going
to show that this sequence does indeed converge, by capturing its limit in a nested
sequence of closed intervals with a size that tends to zero.

We let P_n, Q_n be the relatively prime integers with the ratio c_n; that is

$$\frac{P_n}{Q_n} = a_0 + \cfrac{1}{a_1 + \cfrac{\ddots}{\quad + \cfrac{1}{a_n}}} \qquad \text{for positive integers } a_0, a_1, \ldots, a_n.$$

In particular, $P_0 = a_0$, $Q_0 = 1$ and $P_1 = a_0 a_1 + 1$, $Q_1 = a_1$. For $n > 2$ we will
express P_n in terms of P_{n-1} and P_{n-2}, and Q_n in terms of Q_{n-1} and Q_{n-2}, by simple
recurrence relations which we prove by induction on n.

These relations will enable us to inductively prove various properties of the
fractions P_n/Q_n, and thereby explain their convergence.

Recurrence relations for the convergents. *If P_n and Q_n are the relatively prime
integers such that*

$$\frac{P_n}{Q_n} = a_0 + \cfrac{1}{a_1 + \cfrac{\ddots}{\quad + \cfrac{1}{a_n}}} \qquad \textit{for positive integers } a_0, a_1, a_2, \ldots,$$

then

$$P_0 = a_0, \quad P_1 = a_0 a_1 + 1, \quad and \quad P_n = a_n P_{n-1} + P_{n-2} \ for \ n > 2;$$

$$Q_0 = 1, \quad Q_1 = a_1, \quad and \quad Q_n = a_n Q_{n-1} + Q_{n-2} \ for \ n > 2.$$

Proof. The values of P_0, P_1, Q_0, Q_1 are easy to check, and it is almost as easy to check the recurrence relations for $n = 2$ (exercise). Now suppose that the relations hold for $n = m - 1$, that is, for any sequence of m positive integers $a_0, a_1, \ldots, a_{m-1}$. To prove them for all n, by induction, it suffices to show that they hold for $n = m$.

To do this we first define relatively prime integers P'_j, Q'_j by

$$\frac{P'_j}{Q'_j} = a_1 + \cfrac{1}{a_2 + \cfrac{1}{\ddots + \cfrac{1}{a_{j+1}}}} \qquad \text{for } j = 0, 1, 2, \cdots.$$

Since the recurrences are supposed to hold for *any* sequence of m positive integers, including $a_1, a_2, \ldots, a_m$, we have

$$P'_m = a_m P'_{m-1} + P'_{m-2} \quad and \quad Q'_m = a_m Q'_{m-1} + Q'_{m-2}. \qquad (*)$$

Now the relation between the P/Q fractions and the P'/Q' fractions is

$$\frac{P_j}{Q_j} = a_0 + \frac{1}{P'_j/Q'_j} = a_0 + \frac{Q'_j}{P'_j} = \frac{a_0 P'_j + Q'_j}{P'_j},$$

and we notice (using the Euclidean algorithm) that

$$\gcd(a_0 P'_j + Q'_j, P'_j) = \gcd(P'_j, Q'_j) = 1,$$

so

$$P_j = a_0 P'_j + Q'_j \quad and \quad Q_j = P'_j. \qquad (**)$$

Taking $j = m$ in (**), then applying (*), gives

$$P_m = a_0 P'_m + Q'_m$$

$$= a_0(a_m P'_{m-1} + P'_{m-2}) + a_m Q'_{m-1} + Q'_{m-2}$$

$$= a_m(a_0 P'_{m-1} + Q'_{m-1}) + a_0 P'_{m-2} + Q'_{m-2},$$

$$Q_m = P'_m$$
$$= a_m P'_{m-1} + P'_{m-2}. \tag{***}$$

Also, taking $j = m - 1$ and $j = m - 2$ in (**) gives

$$P_{m-1} = a_0 P'_{m-1} + Q'_{m-1} \quad \text{and} \quad Q_{m-1} = P'_{m-1}$$
$$P_{m-2} = a_0 P'_{m-2} + Q'_{m-2} \quad \text{and} \quad Q_{m-2} = P'_{m-2}.$$

The latter equations allow us to replace all the primed terms in (***) and they become the required recurrence relations for $n = m$:

$$P_m = a_m P_{m-1} + P_{m-2} \quad \text{and} \quad Q_m = a_m Q_{m-1} + Q_{m-2}. \qquad \square$$

From the recurrence relations we quickly obtain some properties of the integers P_n, Q_n that enable us to prove that the convergents P_n/Q_n indeed converge.

1. Since $Q_0 = 1$, $Q_1 = a_1$, $Q_n = a_n Q_{n-1} + Q_{n-2}$, and $a_0, a_1, a_2, \ldots$ are positive integers, it follows by an easy induction that Q_n grows with n and hence $Q_n \geq n$.
2. Another induction (exercise) shows that $P_n Q_{n-1} - Q_n P_{n-1} = (-1)^{n-1}$, whence it follows that

$$\frac{P_n}{Q_n} - \frac{P_{n-1}}{Q_{n-1}} = \frac{(-1)^{n-1}}{Q_n Q_{n-1}}.$$

3. This implies that $a_0 < \frac{P_2}{Q_2} < \frac{P_4}{Q_4} < \frac{P_6}{Q_6} < \cdots < \frac{P_5}{Q_5} < \frac{P_3}{Q_3} < \frac{P_1}{Q_1} \leq a_0 + 1$ and (because of 1)

$$\left| \frac{P_n}{Q_n} - \frac{P_{n-1}}{Q_{n-1}} \right| \leq \frac{1}{n(n-1)}.$$

The last of these results shows that the closed intervals bounded by P_n/Q_n and P_{n-1}/Q_{n-1} are nested and of length tending to 0. Thus they have a unique common point, $\lim_{n \to \infty} P_n/Q_n$, which is the value of the infinite continued fraction

$$a_0 + \cfrac{1}{a_1 + \cfrac{1}{a_2 + \cfrac{1}{\ddots}}}.$$

Exercises

2.8.1 Show that

$$a_0 + \cfrac{1}{a_1 + \cfrac{1}{a_2}} = \frac{a_0(a_1 a_2 + 1) + a_2}{a_1 a_2 + 1},$$

and use the Euclidean algorithm to show that the numerator and denominator on the right are relatively prime.

2.8.2 Conclude from Exercise 2.8.1 that $P_2 = a_0(a_1 a_2 + 1) + a_2$ and $Q_2 = a_1 a_2 + 1$, and deduce that

$$P_2 = a_2 P_1 + P_0 \quad \text{and} \quad Q_2 = a_2 Q_1 + Q_0.$$

2.8.3 Prove $P_n Q_{n-1} - Q_n P_{n-1} = (-1)^{n-1}$ by induction on n.

Two interesting special cases are the continued fractions

$$\cfrac{1}{1 + \cfrac{1}{1 + \cfrac{1}{1 + \cfrac{1}{\ddots}}}} \quad \text{and} \quad \cfrac{1}{2 + \cfrac{1}{2 + \cfrac{1}{2 + \cfrac{1}{\ddots}}}},$$

which represent the numbers $\frac{\sqrt{5}-1}{2}$ and $\sqrt{2} - 1$, respectively.

2.8.4 Use the recurrence relations to show that the convergents for $\frac{\sqrt{5}-1}{2}$ are ratios of successive Fibonacci numbers.

2.8.5 Show that the convergents for $\sqrt{2} - 1$ are ratios of successive terms of the sequence $1, 2, 5, 12, 29, 70, 169, \ldots$, in which each term is twice the previous term plus the term before that.

2.8.6 From Exercise 2.8.5 deduce a result about successive convergents for $\sqrt{2}$.

2.9   Historical Remarks

A slogan to sum up this chapter might be: the basic theory of $\mathbb{R}$ equals Greek mathematics + Infinity (or even Euclid + Infinity). The ancient Greeks gave us integral and rational numbers and the principle of induction by which their properties may be proved. They also gave us infinite processes for approaching irrational numbers, though they did not dare to complete them or "take them to the limit." Thus Euclid, in his *Elements*, Book X, Proposition 2, gave nontermination of the Euclidean algorithm as a criterion for irrationality. He gave no example at this point, but his Proposition 5 of Book XIII immediately implies periodicity, and hence nontermination, of the Euclidean algorithm on the pair $\frac{1+\sqrt{5}}{2}, 1$. This leads us to the continued fraction representation

$$\frac{1 + \sqrt{5}}{2} = 1 + \cfrac{1}{1 + \cfrac{1}{1 + \cfrac{1}{\ddots}}}$$

but it would not have been accepted by the Greeks, since it implies "completing" a process that does not end. The Greeks accepted the "potential" infinity of a process, but not the "actual" infinity of its completion.

Our use of the word "completion" to describe the creation of $\mathbb{R}$ from $\mathbb{Q}$ is appropriate, because it involves the simultaneous completion of infinitely many infinite processes. $\mathbb{R}$ turns out to be a perfect example of an actual infinity, because there is in fact no way to view it as a "potential infinity." $\mathbb{R}$ must be comprehended in its totality or not at all. This remarkable discovery will be explained in Chap. 3, along with the contrasting discovery that $\mathbb{Q}$ *can* be viewed as a "potential infinity."

This was not known when Dedekind discovered the completion of $\mathbb{Q}$ by means of his cuts in 1858, but he was aware of the revolutionary nature of his discovery. In the first publication of his theory, he described the circumstances as follows.

As a professor at the Polytechnic School in Zürich I found myself for the first time obliged to lecture upon the elements of the differential calculus and felt more keenly than ever the lack of a really scientific foundation for arithmetic. In discussing the notion of the approach of a variable magnitude to a fixed limiting value, and especially in proving the theorem that every magnitude that grows continually, but not beyond all limits, must certainly approach a limiting value, I had recourse to geometric evidences. ... that this form of introduction into the differential calculus can make no claim to being scientific, no one will deny. For myself this feeling of dissatisfaction was so overpowering that I made the fixed resolve to keep meditating on the question till I should find a purely arithmetic and perfectly rigorous foundation ... I succeeded Nov. 24, 1858

Dedekind (1872), pp. 1–2

Dedekind's desire to avoid "recourse to geometric evidences" in favor of a "purely arithmetic" foundation was part of a nineteenth century movement away from geometric foundations in mathematics. As we saw in Chap. 1, the Greeks took the discovery of irrational quantities to mean that geometric magnitudes are more extensive than numbers, and for that reason they favored geometry as the foundation of mathematics. This attitude prevailed until the nineteenth century, partly because there was as yet no arithmetic model of the line. However, confidence in geometry was weakened by the discovery of non-Euclidean geometry in the 1820s, and the desire for arithmetic foundations was correspondingly strengthened.

The creation of the number line by completion of $\mathbb{Q}$ was a big step towards an arithmetic foundation for analysis, but further digging was required. $\mathbb{Q}$ itself lacked a proper foundation as long as basic results, such as $ab = ba$, were justified by appeal to geometric intuition. Truly arithmetic proofs of the basic results had to wait for the method of induction to mature beyond the sporadic descent arguments that occur in Euclid.

Fig. 2.4 Hermann Grassmann and Richard Dedekind

The first, rough, idea of proving properties of numbers by establishing them for 1 and working upwards seems to occur in the work of Levi ben Gershon (1321). He used the idea to prove the basic formulas for permutations and combinations, such as the fact that there are $n!$ permutations of n things. Induction proofs in almost the modern ascent format—a *base step* that establishes a property for $n = 1$ (or some other initial value), and an *induction step* showing that the property propagates from n to $n + 1$—occur in Pascal (1654), a book that introduced the so-called "Pascal's triangle" to European readers.

By the nineteenth century, this form of induction was in common use, but it took a brilliant mathematical outsider to see that induction was the absolute foundation of arithmetic. This was the message of the Grassmann (1861) inductive proof of the ring properties of the integers, though it went unnoticed by most mathematicians. The rediscovery of Grassmann's results by Dedekind (1888) and Peano (1889) confirmed the importance and naturalness of his idea. The subsequent development of *set theory*, as we will see in Chap. 6, not only reaffirmed the importance of induction, but also showed that it extends to all kinds of infinity.

2.9.1 $\mathbb{R}$ *as a Complete Ordered Field*

The concept of a complete Archimedean ordered set is motivated by our geometric intuition of the line, which was also the ancient Greek intuition. The Archimedean property, as its name suggests, was mentioned by Archimedes. But before him it was stated by Euclid:

Two unequal magnitudes being set out, if from the greater there be subtracted a magnitude greater than its half, and from that which is left a magnitude greater than its half, and if this process be repeated continually, there will be left some magnitude which will be less than the lesser magnitude set out.

Elements, Book X, Proposition 1.

The concept of field is more modern and of algebraic origin. Although the Greeks essentially knew the field $\mathbb{Q}$, their geometric concept of product did not allow unlimited multiplication of magnitudes, so there was no "field of magnitudes." The concept of field developed in parallel with the development of algebra from the sixteenth century onwards, as mathematicians gradually became conscious of the rules for adding and multiplying numbers and symbolic expressions.

The concept of an *ordered* field, and its specialization to the complete case, was considered by Dedekind (1872). However, it emerged more dramatically from a surprising development of the 1890s: the *geometrization* of algebra. Against the general tide of arithmetization, Hilbert (1899) showed that the nine properties defining a field,

$$a + b = b + a \qquad ab = ba$$

$$a + (b + c) = (a + b) + c \qquad a(bc) = (ab)c$$

$$a + 0 = a \qquad a \cdot 1 = a$$

$$a + (-a) = 0 \qquad a \cdot a^{-1} = 1 \quad \text{for } a \neq 0$$

$$a(b + c) = ab + ac,$$

are equivalent to four geometric axioms. Updating the approach of Euclid, Hilbert introduced undefined objects called "points" and "lines," subject to the following axioms:

1. Through any two points there is a unique line.
2. Any two lines meet in a unique point.
3. There are four points, no three of which lie on the same line.
4. If points A, B, C, D, E, F lie alternately on two lines, then the intersections of the lines AB and DE, BC and EF, CD and FA, lie on a line (shown dashed in Fig. 2.5).

The first three axioms define what is called a *projective plane*, one line of which (chosen arbitrarily) is called the *line at infinity*. The intuition for these three axioms is that the line at infinity is the horizon and that lines meeting on the horizon are parallel. The fourth axiom is the *theorem of Pappus*, so-called because it becomes a theorem when the projective plane is supplied with coordinates. To be precise: in a plane with coordinates, lines have linear equations, and we can compute the intersections of the above lines and show that they lie on a line using the field properties. This is essentially the theorem proved by Pappus of Alexandria around 300 CE.

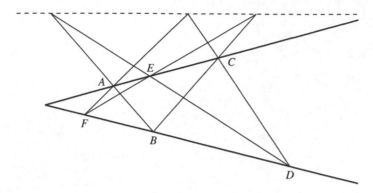

Fig. 2.5 The theorem of Pappus

Conversely, if a projective plane satisfies Axiom 4 we can define coordinates, with sum and product operations, and show that the nine field axioms are satisfied. Thus the part of the complete ordered field concept *not* originating in geometry turns out to have a geometric interpretation: the field concept is captured by the structure of a projective plane satisfying the Pappus theorem. Even more remarkably, the Pappus theorem can be held responsible specifically for $ab = ba$. This is because there is a weaker geometric theorem (implied by Pappus), called the *Desargues theorem*, which implies all of the field properties *except $ab = ba$*.

So, to return to the question raised in Sect. 1.1, it is not unreasonable to seek a geometric explanation of $ab = ba$. But if you want the other field properties, Pappus is a better explanation than Euclid!

Chapter 3
Infinite Sets

PREVIEW

The construction of the continuous set $\mathbb{R}$ by "filling the gaps" in the set $\mathbb{Q}$ of rational numbers seems completely natural and simple, in hindsight. However, there is a huge difference between $\mathbb{Q}$ and $\mathbb{R}$. While both are infinite sets, $\mathbb{R}$ is "more infinite" than $\mathbb{Q}$. The purpose of the present chapter is to explain precisely what this means, and to compare some other important infinite sets with $\mathbb{Q}$ and $\mathbb{R}$.

The "smallest" infinite sets are those called *countably* infinite. The definitive example is the set of positive integers $\mathbb{N} = \{1, 2, 3, 4, \ldots\}$. A set is called *countable* if its members can be arranged in a (possibly infinite) list: 1st member, 2nd member, 3rd member, $\ldots$, so that each member occurs at some positive integer position. Perhaps surprisingly, the rational numbers can be arranged in such a list, so $\mathbb{Q}$ is countably infinite. In fact, any set with members that have "finite descriptions" in some reasonable sense turns out to be countable.

This is not the case for the real numbers, many of which require infinite descriptions, such as infinite decimals. Indeed, there are some dramatic proofs that $\mathbb{R}$ is *not* countable. We give a couple of these uncountability proofs, and also find several sets that are "equinumerous" with $\mathbb{R}$, and hence also uncountable.

The uncountability of $\mathbb{R}$ leads us to expect some difficulties and surprises when we come to investigate sets of real numbers. To prepare for what is in store, we prove some classical theorems about sets of real numbers and introduce a notable example—the so-called *Cantor set*.

3.1 Countably Infinite Sets

A set is said to be *countable* if its members can be enumerated—first member, second member, third member, and so on—and each member eventually appears in the enumeration. Notice that we do not assume that the enumeration ever comes to an end. If it does not, the set is called *countably infinite*.

J. Stillwell, *The Real Numbers: An Introduction to Set Theory and Analysis*,
Undergraduate Texts in Mathematics, DOI 10.1007/978-3-319-01577-4_3,
© Springer International Publishing Switzerland 2013

The definitive example of a countably infinite set is the set of positive integers

$$\mathbb{N} = \{1, 2, 3, 4, 5, \ldots\}.$$

Another way to say that a set S is countably infinite is to say that the members of S can be put in one-to-one correspondence with the members of $\mathbb{N}$: the first member of S corresponds to 1, the second member to 2, and so on. A little less formally, a set is countably infinite if its members can be arranged in an infinite list:

first member, second member, third member, ...

There are many examples of countably infinite sets, the most important of which are the following

1. Any infinite subset of $\mathbb{N}$, because (by induction) any such subset has a least member, second least member, third least member, and so on indefinitely. Examples are {even numbers}, {squares}, and {primes}.
2. The set of integers, $\mathbb{Z} = \{\ldots, -3, -2, -1, 0, 1, 2, 3, \ldots\}$. $\mathbb{Z}$ is countable because its members can be listed as follows:

$$0, \ 1, \ -1, 2, \ -2, \ 3, \ -3, \ \ldots$$

(That is, begin with 0 and then alternate members of $\mathbb{N}$ with their negatives.)
3. The set of rational numbers between 0 and 1. These numbers can be arranged in the following list:

$$\frac{1}{2}; \ \frac{1}{3}, \frac{2}{3}; \ \frac{1}{4}, \frac{3}{4}; \ \frac{1}{5}, \frac{2}{5}, \frac{3}{5}, \frac{4}{5}; \ \frac{1}{6}, \frac{5}{6}; \ \cdots$$

(That is, we first list the fractions with denominator 2, then those with denominator 3, then those with denominator 4, and so on, including only those fractions that are in lowest terms.)
4. The set $\mathbb{Q}^+$ of positive rational numbers. To list the members of this set we group positive fractions according to the *sum* of their numerator and denominator: first those with sum equal to 2, then those with sum equal to 3, and so on. Within each group we list the fractions in increasing order of their numerators, again including only fractions in lowest terms. Then the list begins

$$\frac{1}{1}; \ \frac{1}{2}, \frac{2}{1}; \ \frac{1}{3}, \frac{3}{1}; \ \frac{1}{4}, \frac{2}{3}, \frac{3}{2}, \frac{4}{1}; \ \frac{1}{5}, \frac{5}{1}; \ \frac{1}{6}, \frac{2}{5}, \frac{3}{4}, \frac{4}{3}, \frac{5}{2}, \frac{6}{1}; \ \cdots$$

5. The set $\mathbb{Q}$ of all rational numbers. To list the members of $\mathbb{Q}$ we first list 0, then alternate members of $\mathbb{Q}$ with their negatives (the same trick we used to enumerate $\mathbb{Z}$, given the enumeration of $\mathbb{N}$):

$$0; \ \frac{1}{1}, -\frac{1}{1}; \ \frac{1}{2}, -\frac{1}{2}, \frac{2}{1}, -\frac{2}{1}; \ \frac{1}{3}, -\frac{1}{3}, \frac{3}{1}, -\frac{3}{1}; \ \cdots$$

6. The set $\overline{\mathbb{Q}}$ of all algebraic numbers, where an *algebraic number* is a root of a polynomial equation with integer coefficients. A polynomial of degree n,

$$a_n x^n + a_{n-1} x^{n-1} + \cdots + a_1 x + a_0 = 0, \qquad (*)$$

has at most n distinct roots, so the main problem in enumerating algebraic numbers is to enumerate all the polynomial equations (*), where $a_0, a_1, \ldots, a_n$ are integers.

To do this we consider the number

$$h = n + |a_n| + |a_{n-1}| + \cdots + |a_1| + |a_0|,$$

which is called the *height* of the equation (*). There are only a finite number of equations of height $\leq h$, so we can make a list of all equations (*) with integer coefficients by first listing those of height 1, then those of height 2, and so on.

Then if we list, along with each equation, its finitely many roots, we obtain a list of all algebraic numbers.

7. The set of all finite subsets of $\mathbb{N}$. A list of finite subsets may be constructed inductively as follows. At stage zero, list the empty set, $\emptyset$. Then, assuming all subsets of $\{1, 2, \ldots, n\}$ have been listed by stage n, at stage $n + 1$ list all sets obtained by inserting the number $n + 1$ in previous sets. Then all subsets of $\{1, 2, \ldots, n, n + 1\}$ have been listed by the end of stage $n + 1$. The list therefore looks like this:

$$\emptyset, \{1\}, \{2\}, \{1, 2\}, \{3\}, \{1, 3\}, \{2, 3\}, \{1, 2, 3\}, \ldots$$

8. The set $\mathbb{N}^{<\omega}$ of all finite sequences of positive integers. This seems like listing finite subsets, except that order is important and elements may be repeated. So it is more like the listing of polynomials above, and indeed we can use a similar concept of "height." We assign the n-element sequence $\langle a_1, a_2, \ldots, a_n \rangle$ the height

$$n + a_1 + \cdots + a_n.$$

Then there are only a finite number of sequences of given height, so we can list all sequences by listing those of height 1, then those of height 2, and so on.

As special cases, the sets of all ordered pairs, ordered triples, and so on, of members of a countable set are themselves countable.

3.1.1 The Universal Library

The last example of a countable set above can be interpreted more dramatically. A word, a sentence, even a whole book is nothing but a finite sequence of symbols, which we could encode by natural numbers (indeed we need only finitely many symbols if we include all the letters of the alphabet, punctuation symbols, and the blank space). Therefore, the list of finite sequences of natural numbers in principle includes every book that has been, or will ever be, written. This leads to the idea of a *universal library*, which has been a plaything for several writers. One of the most eloquent was the set theorist (and sometime dramatist) Felix Hausdorff, who wrote in his *Grundzüge der Mengenlehre* of 1914, pp. 61–62 (my translation):

> If one adds to the letters further elements such as punctuation marks, spaces, numerals, notes, etc., then one sees that the set of all books, catalogs, symphonies, and operas is countable, and it remains countable when one allows countably many symbols (but only finitely many for each work). On the other hand, if one confines oneself to a finite number of symbols, and to works of a bounded length, say by allowing words no longer than one hundred letters and books of no more than one million words, then the set is finite. And if one supposes, with Giordano Bruno, an infinite numbers of worlds with speaking, writing, and music-making inhabitants, then it follows with mathematical certainty that in infinitely many of these worlds the same opera, with the same libretto, by a composer, librettist, conductor, and singers with the same names, will be performed.

One might add that, if music is digitized, so that each performance becomes a finite sequence of bits, then the set of performances is countable and there is a *universal music library*. However, a more interesting question is: how big must the universal music library be if music is *not* digitized? A similar question is: how big must a library be to hold all possible handwritten manuscripts? We take up these questions in Sect. 3.4.

Exercises

An amusing model of countability, which apparently first appeared in the book *One, Two, Three, ... Infinity* of Gamow (1947), is called *Hilbert's hotel*. Hilbert's hotel has a countable infinity of rooms—room 1, room 2, room 3, and so on—and sets are counted by packing them into Hilbert's hotel, one member per room. Thus, the set $\mathbb{N}$ fills Hilbert's hotel as shown in Fig. 3.1.

Even though the hotel is full, it can make room for a new guest, say 0, by having each occupant move into the next room, as shown in Fig. 3.2.

3.1.1 Explain how to make room for a countable infinity of new guests, such as $-1, -2, -3, -4, \ldots$.

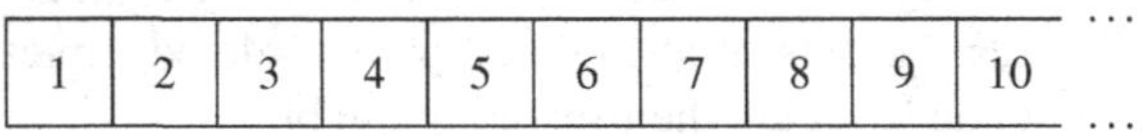

| 1 | 2 | 3 | 4 | 5 | 6 | 7 | 8 | 9 | 10 | ... |

Fig. 3.1 Hilbert's hotel occupied by the members of $\mathbb{N}$

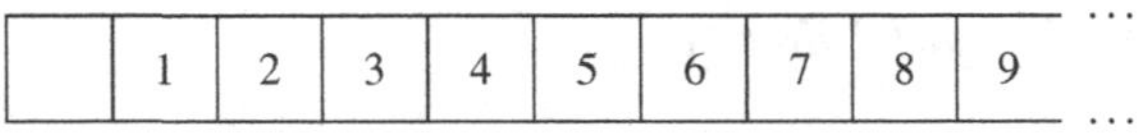

| | 1 | 2 | 3 | 4 | 5 | 6 | 7 | 8 | 9 | ... |

Fig. 3.2 Making room for one more

Fig. 3.3 Rooms for the first busload

Now suppose that infinitely many buses arrive at the hotel, each carrying a countable infinity of passengers. Suppose that the passengers $b_1, b_2, b_3, \ldots$ of the first bus are given rooms as shown in Fig. 3.3.

That is, skip one room after the first passenger, skip two rooms after the second passenger, and so on.

3.1.2 Find rooms for the second busload, in such a way that there remain blocks of $1, 2, 3, \ldots$ empty rooms.

3.1.3 Deduce that it is possible to accommodate all passengers from all buses, so as to exactly fill Hilbert's hotel.

3.2 An Explicit Bijection Between $\mathbb{N}$ and $\mathbb{N}^2$

In the flurry of results in the previous section, we skimmed over one that is important enough to study in some detail: a bijection between the set $\mathbb{N}$ of positive integers and the set $\mathbb{N}^2$ of *ordered pairs* of positive integers. This bijection is crucial to several other bijections that we construct later, so it is important to be aware of it. And, indeed, the bijection between $\mathbb{N}$ and $\mathbb{N}^2$ can be made very clear, both pictorially and by means of a simple quadratic function. We begin with a picture.

Figure 3.4 shows the points $\langle m, n \rangle$ of $\mathbb{N}^2$ in their usual grid arrangement, for small values of m and n. Also shown on the grid is a series of diagonal dotted lines that show how to enumerate all the points in $\mathbb{N}^2$.

We take $\langle 1, 1 \rangle$ as point number 1 on the list, then continue numbering points as $2, 3$ on the first diagonal, then $4, 5, 6$ on the next diagonal, and so on. The numbers of the points are shown in gray. It is clear that each point eventually gets a number with this scheme, so we have established a bijection between $\mathbb{N}$ and $\mathbb{N}^2$.

Moreover, we can obtain a formula for the number of point $\langle m, n \rangle$ as follows. The point $\langle k - 1, 1 \rangle$ at the end of the $(k - 2)$nd diagonal obviously has number

$$1 + 2 + 3 + \cdots + (k - 1) = k(k - 1)/2.$$

One more step brings us to the point $\langle 1, k \rangle$ at the beginning of the $(k - 1)$st diagonal, and another $m - 1$ steps along this diagonal brings us to the point $\langle m, k - m + 1 \rangle$. Thus,

$$\text{number of point } \langle m, k - m + 1 \rangle = m + k(k - 1)/2.$$

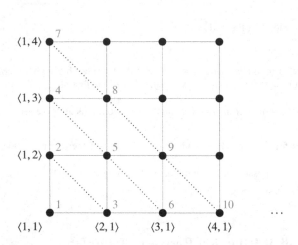

Fig. 3.4 Enumerating the points in $\mathbb{N}^2$

Rewriting this formula in terms of $n = k - m + 1$, so $k = m + n - 1$, we get

$$\text{number of point } \langle m, n \rangle = m + (m + n - 1)(m + n - 2)/2.$$

We will denote the number of point $\langle m, n \rangle$ by $p(m, n)$ (so you can think of p standing for "pairing").

Exercises

3.2.1 Find a polynomial bijection from $\mathbb{N}^3$ to $\mathbb{N}$. What is its degree?

3.2.2 Explain why there are countably many points in $\mathbb{R}^2$ with rational coordinates.

3.2.3 Show that there are countably many circles in $\mathbb{R}^2$ with rational center and rational radius.

3.2.4 Also, show that there are countably many circles with three rational points.

3.3 Sets Equinumerous with $\mathbb{R}$

The countable sets studied in Sect. 3.1—$\mathbb{Z}$, $\mathbb{Q}$, and $\overline{\mathbb{Q}}$—lie more and more densely on the number line $\mathbb{R}$, yet no method is apparent for listing all the members of $\mathbb{R}$. The problem appears to be that the description of a real number is generally infinite, and the methods above are good only for listing objects with finite descriptions. Before attempting to prove that $\mathbb{R}$ is *not* countable, however, we will look at some interesting sets that are *equinumerous* with $\mathbb{R}$.

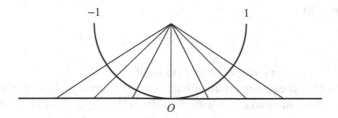

Fig. 3.5 Bijection between an interval and the line

Definition. Sets are *equinumerous*, or *of the same cardinality*, if there is a one-to-one correspondence (bijection) between their elements.

Thus, the countably infinite sets in the previous section are equinumerous with $\mathbb{N}$, or of the same cardinality as $\mathbb{N}$. The sets equinumerous with $\mathbb{R}$ include some subsets of $\mathbb{R}$, some sets that contain $\mathbb{R}$, and also some sets consisting of infinite objects derived from the countable set $\mathbb{N}$. We sometimes say that such sets have *continuum-many* elements.

1. The first example of a subset equinumerous with $\mathbb{R}$ is the open interval

$$(-1, 1) = \{x \in \mathbb{R} : -1 < x < 1\}.$$

An easy way to see a bijection between $(-1, 1)$ and $\mathbb{R}$ is to imagine $(-1, 1)$ bent into a semicircle that rests on the number line at O, as shown in Fig. 3.5. Rays from the center of the semicircle establish a one-to-one correspondence between points of $(-1, 1)$ and points of the line.

A similar picture shows that *any* open interval

$$(a, b) = \{x \in \mathbb{R} : a < x < b\},$$

where $a < b$, is equinumerous with the whole number line. An interesting case, not needing the picture, is the interval $(-\pi/2, \pi/2)$. The tan function maps this interval one-to-one onto $\mathbb{R}$, as one can see from the graph of the tan function.

2. The closed interval $[0, 1] = \{x \in \mathbb{R} : 0 \leq x \leq 1\}$ is equinumerous with the open interval $(0, 1)$, and hence with $\mathbb{R}$. We show this by constructing a bijection of $[0, 1]$ onto $(0, 1)$. Consider the numbers

$$r_1 = \frac{1}{2}, \quad r_2 = \frac{3}{4}, \quad r_3 = \frac{7}{8}, \quad r_4 = \frac{15}{16}, \quad \ldots,$$

which belong to both $[0, 1]$ and $(0, 1)$. To map $[0, 1]$ one-to-one onto $(0, 1)$ we send

$$0, \ 1, \ r_1, \ r_2, \ r_3, \ \ldots \quad \text{in} \quad [0, 1]$$

respectively to

$$r_1, \ r_2, \ r_3, \ r_4, \ r_5, \ \ldots \quad \text{in} \quad (0, 1),$$

and send every other member of $[0, 1]$ to itself.

Similarly, $\mathbb{R}$ is equinumerous with any closed interval $[a, b]$ with $a < b$.

3. Another subset equinumerous with $\mathbb{R}$ is the set[1] $\mathbb{R} - \mathbb{Q}$ of irrational numbers. As we saw in the previous section, the rational numbers can be enumerated $r_1, r_2, r_3, \ldots$. From this list we can construct a countable infinity of irrational numbers, for example

$$s_1 = \sqrt{2} r_1, \quad s_2 = \sqrt{2} r_2, \quad s_3 = \sqrt{2} r_3, \quad \ldots$$

We can now define a one-to-one function from $\mathbb{R}$ onto $\mathbb{R} - \mathbb{Q}$ by sending

$$r_1, \ s_1 \ , r_2, \ s_2, \ r_3, \ s_3, \ \ldots \quad \text{in} \quad \mathbb{R}$$

respectively to

$$s_1, \ s_2, \ s_3, \ s_4, \ s_5, \ s_6, \ \ldots \quad \text{in} \quad \mathbb{R} - \mathbb{Q}$$

and sending each other member of $\mathbb{R}$ to itself.

4. The set $\mathcal{P}(\mathbb{N})$ of subsets of $\mathbb{N}$. (The letter $\mathcal{P}$ stands for "power set" and means "all subsets of." We say more about this operation in Chap. 6.)

A subset S of $\mathbb{N}$ can be described by an infinite sequence of 0s and 1s, with 1 in the nth place if and only if $n \in S$. Such a sequence can be interpreted as the binary expansion of a number in $[0, 1]$. The only problem is that different subsets can give the same number, for example

$\{1\}$ is described by $10000\ldots$, which gives the number $0.10000\ldots = \frac{1}{2}$.

$\{2, 3, 4, \ldots\}$ is described by $01111\ldots$, which gives the number $0.01111\ldots = \frac{1}{2}$.

The numbers that correspond to different sets are the *binary fractions* $m/2^n$, and the corresponding subsets of $\mathbb{N}$ are the finite sets and their complements. Leaving these exceptional numbers and sets aside for the moment, we have a bijection from

$$[0, 1] - \{\text{binary fractions}\} \quad \text{onto}$$

$$\mathcal{P}(\mathbb{N}) - \{\text{finite sets and their complements}\}.$$

---

[1] In this book we use an ordinary minus sign to denote set difference. This is convenient later to show the parallel between set difference and number difference in measure theory. In any case, it will always be clear what kind of objects we are taking the difference of.

Finally, by a bijection between the countable sets

{binary fractions}   and   {finite sets and their complements}

gives a bijection of $[0, 1]$ onto $\mathcal{P}(\mathbb{N})$. So $\mathcal{P}(\mathbb{N})$ is equinumerous with $[0, 1]$, and hence with $\mathbb{R}$.

5. The set $\mathbb{N} \times \mathbb{N} \times \mathbb{N} \times \cdots = \mathbb{N}^{\mathbb{N}}$ of infinite sequences of positive integers.

   Each infinite sequence[2] $\langle n_1, n_2, n_3, \ldots \rangle$ of positive integers gives an irrational number

$$x = \cfrac{1}{n_1 + \cfrac{1}{n_2 + \cfrac{1}{n_3 + \cfrac{1}{\ddots}}}}$$

between 0 and 1, because each rational has a finite continued fraction, as we saw in Sect. 2.7. Conversely, each irrational number between 0 and 1 has a continued fraction of the above form, and hence gives an infinite sequence of positive integers.

   Thus, we immediately have a bijection between $\mathbb{N}^{\mathbb{N}}$ and the irrational numbers in $(0, 1)$. The latter set is equinumerous with $(0, 1)$, and hence with $\mathbb{R}$, by an argument like that used to show that $\mathbb{R} - \mathbb{Q}$ is equinumerous with $\mathbb{R}$.

Exercises

3.3.1   Show that $(0, 1) \cup \{1\}$, $(0, 1) \cup \{1, 2\}$, and $(0, 1) \cup \{1, 2, 3\}$ are equinumerous with $\mathbb{R}$. (*Hint*: Build a copy of "Hilbert's Hotel" inside $(0, 1)$.)

3.3.2   Show that $(0, 1) \cup \{1, 2, 3, \ldots\}$ is equinumerous with $\mathbb{R}$.

3.3.3   Show that $(0, 1) \cup (2, 3)$ is equinumerous with $\mathbb{R}$.

3.3.4   Encode each $\langle n_1, n_2, n_3, \ldots \rangle \in \mathbb{N}^{\mathbb{N}}$ by an infinite sequence of 0s and 1s with infinitely many 0s, and hence give another proof that $\mathbb{N}^{\mathbb{N}}$ is equinumerous with $\mathbb{R}$.

3.4   The Cantor–Schröder–Bernstein Theorem

The bijections in the previous section call for a certain amount of ingenuity, in order to construct bijections from maps that are not *quite* bijections. To avoid

---

[2] We use the notation $\langle a, b, c, \ldots \rangle$ for the infinite sequence in conformity with the notations $\langle a, b \rangle$ and $\langle a, b, c \rangle$ for ordered pairs and triples.

ever-increasing demands on our ingenuity in the future, we now prove a theorem that guarantees a bijection when we "almost" have one, namely, when we have a bijection from A to a subset of B, and one from B to a subset of A (or, equivalently, an *injection* from A into B, and one from B into A).

Cantor-Schröder-Bernstein Theorem. *If there are injections $f : A \to B$ and $g : B \to A$ then there is a bijection $h : A \to B$.*

Proof. Consider chains of elements, alternately in A and B, that are connected by alternate applications of f and g. A portion of a chain looks like this,

$$\cdots \overset{g}{\mapsto} a_0 \overset{f}{\mapsto} b_0 \overset{g}{\mapsto} a_1 \overset{f}{\mapsto} b_1 \overset{g}{\mapsto} a_2 \overset{f}{\mapsto} b_2 \overset{g}{\mapsto} \cdots$$

where $\ldots, a_0, a_1, a_2, \ldots \in A$ and $\ldots, b_0, b_1, b_2, \ldots \in B$ and $f(a_k) = b_k$, $g(b_k) = a_{k+1}$. Since f and g are functions on A and B, respectively, each chain extends indefinitely (and uniquely) to the right, possibly repeating the same finite sequence over and over. Since f and g are injective, each chain extends uniquely to the left too, though it may terminate—either at an a that is not in the range of g or at a b that is not in the range of f.

The injectiveness of f and g also implies that any two chains with a common element a_k (or b_k) are identical. Hence distinct chains contain disjoint subsets of A and disjoint subsets of B. This enables us to set up a bijection $h : A \to B$ by piecing together the obvious bijections within each chain.

1. For any a_k in a chain without an initial element, let $h(a_k) = b_k$, in which case $h^{-1}(b_k) = a_k$.
2. For any a_k in a chain with initial element in A, again let $h(a_k) = b_k$, in which case $h^{-1}(b_k) = a_k$.
3. For any chain with initial element in B, let $h(a_k) = b_{k-1}$, in which case $h^{-1}(b_{k-1}) = a_k$.

It follows, from the partitioning of elements of A and B among the chains, that each $a \in A$ is now paired with a unique $h(a) \in B$, and each $b \in B$ is paired with a unique $h^{-1}(b) \in A$. $\square$

3.4.1 More Sets Equinumerous with $\mathbb{R}$

Armed with the Cantor–Schröder–Bernstein theorem, we can now exhibit some spectacular examples of sets equinumerous with $\mathbb{R}$. The main problem that Cantor-Schröder-Bernstein has to overcome is the *ambiguity* of decimal expansions. That is, the same number is sometimes represented by two different decimal expansions; for example, $1/2$ is represented by both 0.5 and $0.49999\ldots$. It will be seen below that we can get around this problem by constructing injections and applying Cantor–Schröder–Bernstein, rather than attempting to construct bijections directly. However, the ambiguity of decimal expansions continues to cause trouble in the future, and we will eventually work with a different set of continuum cardinality—the set $\mathbb{N}^{\mathbb{N}}$ introduced in Sect. 3.3—to avoid such difficulties.

The first example is the plane $\mathbb{R}^2$. When this example was discovered by Cantor in 1877 he wrote to Dedekind: "I see it but I don't believe it," apparently astonished that a two-dimensional continuum of points could be equinumerous with the one-dimensional continuum. (See Gouvêa 2011 for an engaging account of this episode.) We continue the numbering of examples from the previous section.

6. The set $\mathbb{R}^2$ of *ordered pairs* $\langle x, y \rangle$ of real numbers.

We have an obvious injection of $\mathbb{R}$ into $\mathbb{R}^2$—namely, send each $x \in \mathbb{R}$ to the ordered pair $\langle x, 0 \rangle$—so it remains to find an injection of $\mathbb{R}^2$ into $\mathbb{R}$. We inject $\mathbb{R}^2$ in two stages.

First, map $\mathbb{R}^2$ bijectively onto the "open unit square" $(0, 1)^2$ by mapping each $\mathbb{R}$ bijectively onto the open interval $(0, 1)$ as in example 1 of the previous section.

Now take any $\langle x, y \rangle \in (0, 1)^2$ and consider the decimal expansions of x and y:

$$x = 0.a_1 a_2 a_3 \ldots, \quad y = 0.b_1 b_2 b_3 \ldots$$

We can choose these decimal expansions uniquely by not allowing expansions ending with $999\ldots$. Hence we can send each pair to the well-defined decimal number

$$z = 0.a_1 b_1 a_2 b_2 a_3 b_3 \ldots.$$

Moreover, since z cannot end with $999\ldots$ and hence equal another decimal expansion (because neither x nor y do), different pairs $\langle x, y \rangle$ give different numbers z. Thus, the map $\langle x, y \rangle \mapsto z$ is an injection of $(0, 1)^2$ into $\mathbb{R}$. Combining this injection with the bijection $\mathbb{R}^2 \to (0, 1)^2$ gives an injection $\mathbb{R}^2 \to \mathbb{R}$, as required.

7. The set $\mathbb{R} \times \mathbb{R} \times \mathbb{R} \times \cdots = \mathbb{R}^{\mathbb{N}}$ of sequences $\langle x_1, x_2, x_3, \ldots \rangle$ of real numbers.

Again we have an obvious injection of $\mathbb{R}$ into $\mathbb{R}^{\mathbb{N}}$, by sending x to $\langle x, 0, 0, \ldots \rangle$, and it remains to find an injection of $\mathbb{R}^{\mathbb{N}}$ into $\mathbb{R}$.

The first step is to map $\mathbb{R}^{\mathbb{N}}$ bijectively onto the "infinite-dimensional open unit cube," $(0, 1)^{\mathbb{N}}$, consisting of the sequences $\langle x_1, x_2, x_3, \ldots \rangle$ where each $x_i \in (0, 1)$. As in the previous example, this is done by mapping each $\mathbb{R}$ bijectively onto $(0, 1)$ as in example 1 from the previous section.

Now take any $\langle x_1, x_2, x_3, \ldots \rangle \in (0, 1)^{\mathbb{N}}$ and consider the decimal expansions of $x_1, x_2, x_3, \ldots$:

$$x_1 = 0.a_{11} a_{12} a_{13} \cdots$$

$$x_2 = 0.a_{21} a_{22} a_{23} \cdots$$

$$x_3 = 0.a_{31} a_{32} a_{33} \cdots$$

$$\vdots$$

We choose these expansions uniquely by not allowing any expansion to end with $999\ldots$. We can pack all the decimal digits a_{mn} in this array into the decimal expansion of a single number x by rearranging the array of digits into a list, just as we enumerated the ordered pairs $\langle m, n \rangle$ of natural numbers in Sect. 3.2. The resulting decimal expansion begins

$$x = 0.a_{11}a_{21}a_{12}a_{31}a_{22}a_{13}a_{41}a_{32}a_{23}a_{14}\ldots$$

and in general a_{mn} is in place number $p(m, n)$, where p is the quadratic function of m and n defined in Sect. 3.2.

Each decimal place of x gets filled, so the sequence $\langle x_1, x_2, x_3, \ldots \rangle$ is sent to a well-defined $x \in \mathbb{R}$. Moreover, x cannot end with $999\ldots$, because none of $x_1, x_2, x_3, \ldots$ do, so different sequences give different numbers x.

Thus, the map $\langle x_1, x_2, x_3, \ldots \rangle \mapsto x$ is an injection from $(0, 1)^{\mathbb{N}}$ into $\mathbb{R}$. Combining it with the bijection from $\mathbb{R}^{\mathbb{N}}$ to $(0, 1)^{\mathbb{N}}$ gives an injection $\mathbb{R}^{\mathbb{N}} \to \mathbb{R}$, as required.

8. The set of all continuous functions $f : \mathbb{R} \to \mathbb{R}$.

We will define and study continuous functions carefully in Chap. 4. But for now it is sufficient to know that *a continuous function $f : \mathbb{R} \to \mathbb{R}$ is completely determined by its values on the set $\mathbb{Q}$ of rational numbers*. It follows, since we can list $\mathbb{Q}$ as a sequence $r_1, r_2, r_3, \ldots$, that f is completely determined by the sequence

$$\langle f(r_1), f(r_2), f(r_3), \ldots \rangle \in \mathbb{R}^{\mathbb{N}}.$$

We can therefore inject the set of real continuous functions f into $\mathbb{R}$ by combining the map $f \mapsto \langle f(r_1), f(r_2), f(r_3), \ldots \rangle$ into $\mathbb{R}^{\mathbb{N}}$ with the injection $\mathbb{R}^{\mathbb{N}} \to \mathbb{R}$ found in the previous example.

Conversely, we certainly have an injection of $\mathbb{R}$ into the set of continuous functions. Just send the real number $c \in \mathbb{R}$ to the constant function $f(x) = c$. Thus, it follows from the Cantor–Schröder–Bernstein theorem that the set of continuous functions $f : \mathbb{R} \to \mathbb{R}$ is equinumerous with $\mathbb{R}$.

3.4.2 The Universal Jukebox

Let me begin with a quote from *The Six Gateways of Knowledge*, by Lord Kelvin, an address to the Birmingham and Midland Institute, delivered in the Town Hall, Birmingham, on October 3rd, 1883. It was later published in his *Popular Lectures and Addresses*, Kelvin (1889), volume 1, pp. 274–275.

> But now for what really to me seems a marvel of marvels: think what a complicated thing is the result of an orchestra playing …. Think of the condition of the air, how it is lacerated sometimes in a complicated effect. Think of the smooth gradual increase and diminution of pressure—smooth and gradual though taking place several hundred times in

Fig. 3.6 The universal jukebox

a second—when a piece of beautiful harmony is heard! Whether, however, it be the single note of the most delicate sound of a flute, or the purest piece of harmony of two voices singing perfectly in tune; or whether it be the crash of an orchestra, and the high notes, sometimes even screechings and tearings of the air, which you may hear fluttering above the sound of the chorus—think of all that, and yet .... A single curve, drawn in the manner of the curve of prices of cotton, describes all that the ear can possibly hear, as the result of the most complicated musical performance.

The phenomenon described by Kelvin—the superposition of sound waves into a single wave—had already been exploited by Edison when he first recorded sound in 1877. By using the single sound wave to drive a vibrating needle, Edison transferred the wave onto a wax cylinder, from which the sound could be replayed by reversing the process. Even today, when digitized music is everywhere, many audiophiles prefer the analog sound captured by the continuous wave on the grooves of a vinyl disk. And vinyl disks are still played in old-style jukeboxes.

If we accept that a perfectly faithful sound recording needs to be a continuous function, then the universal jukebox needs to be a repository of all continuous functions. The last example of the previous section shows, amazingly, that each continuous function may be encoded by a single real number. Thus, we can take the universal jukebox to be the number line, with each musical performance represented by a single point (Fig. 3.6). This is *really* the marvel of marvels!

Exercises

Many of our previous results on countable sets and sets equinumerous with $\mathbb{R}$ can be proved more simply with the help of the Cantor–Schröder–Bernstein theorem.

3.4.1 Show that $\langle m, n \rangle \mapsto 2^m 3^n$ gives an injection $\mathbb{N}^2 \to \mathbb{N}$, and hence show that $\mathbb{N}^2$ is equinumerous with $\mathbb{N}$.

3.4.2 If $p_1, p_2, p_3, \ldots$ are the prime numbers, show that the map

$$\langle n_1, n_2, \ldots, n_k \rangle \mapsto 2^{n_1} 3^{n_2} \cdots p_k^{n_k}$$

is an injection of {finite sequences of positive integers} into $\mathbb{N}$. Hence show that {finite sequences of positive integers} is equinumerous with $\mathbb{N}$.

3.4.3 Prove that $[0,1]$ is equinumerous with $(0,1)$ by finding suitable injections from one to the other.

Moreover, many results we previously had not contemplated become easy with the Cantor–Schröder–Bernstein theorem. For example:

3.4.4 For an arbitrary $S \subseteq \mathbb{R}$, show that $(0, 1) \cup S$ is equinumerous with $\mathbb{R}$.

An interesting application of the fact that continuous functions are determined by their values on $\mathbb{Q}$ is the following theorem of Cauchy (1821), pp. 104–106. *If $f : \mathbb{R} \to \mathbb{R}$ is continuous and additive—that is, $f(x + y) = f(x) + f(y)$ for all $x, y \in \mathbb{R}$—then $f(x) = ax$ for some constant a.*

3.4.5 If $f(x + y) = f(x) + f(y)$ and $f(1) = a$, deduce that $f(r) = ra$ for each rational number r.
3.4.6 Deduce from Exercise 3.4.5 that, if f is continuous, then $f(x) = ax$.

3.5 The Uncountability of $\mathbb{R}$

The set $\mathbb{R}$ of real numbers is not a countable set, because we can show that any countable set of real numbers is not all of $\mathbb{R}$. This result shows the need for set theory as a *theory of infinity*, since different kinds of infinity exist. Here are two different (though distantly related) arguments for this famous result.

3.5.1 The Diagonal Argument

The first argument, due to Cantor (1891), takes any countable set of real numbers and explicitly finds a number different from each member of S.

Countable sets do not include all real numbers. *If S is a countable set of real numbers, then there is a member of $[0, 1]$ not in S.*

Proof. Suppose that $S = \{x_1, x_2, x_3, \ldots\}$ is a countable set of real numbers. Each number x_n can be written as an infinite decimal, and we imagine all of these decimal expansions listed in an infinite table, such as the following:

$$
\begin{array}{ll}
x_1 & 0.\underline{1}1111\ldots \\
x_2 & 3.1\underline{4}159\ldots \\
x_3 & 1.23\underline{4}56\ldots \\
x_4 & 0.212\underline{1}2\ldots \\
x_5 & 1.4142\underline{3}\ldots \\
& \;\;\vdots
\end{array}
$$

Ignoring the parts of each number before the decimal point, we construct a number x that differs from each x_n by the simple expedient of *making x different from x_n in the nth decimal place*. To be specific, let

$$
n\text{th decimal place of } x = \begin{cases} 2 & \text{if } n\text{th decimal place of } x_n \text{ is 1} \\ 1 & \text{if } n\text{th decimal place of } x_n \text{ is not 1} \end{cases}
$$

This makes sure that x not only has a different decimal expansion from x_n; x is also a different number, because we have avoided the ambiguous numbers such as

$0.999\ldots = 1.000\ldots$ by using only the digits 1 and 2. For the numbers $x_1, x_2, x_3, \ldots$ tabulated above, we have

$$x = 0.21121\ldots$$

Thus, the countable set S does not include all members of $\mathbb{R}$. Specifically, it does not include the number x in $[0, 1]$.   □

The argument above is known as the *diagonal argument* because it uses the digits on the diagonal of the table of decimal expansions (underlined in the example above). There are many variations of the diagonal argument; indeed, it is hard to get away from it when proving that $\mathbb{R}$ is uncountable.

3.5.2   The Measure Argument

Again we show that a countable set of real numbers cannot fill the interval $[0, 1]$, but this time the argument shows that a countable set falls far short of filling $[0, 1]$.

Countable sets have small measure. *If S is a countable set of real numbers, then a large fraction of the members of $[0, 1]$ are not in S.*

Proof. Suppose that $S = \{x_1, x_2, x_3, \ldots\}$. Suppose that we enclose each number x_n by an open interval U_n of width $1/10^n$. Then the amount of any interval, say $[0, 1]$, covered by $U_1, U_2, U_3, \ldots$ is at most

$$\frac{1}{10} + \frac{1}{10^2} + \frac{1}{10^3} + \cdots = \frac{1}{9}.$$

Thus, not all of $[0, 1]$ is covered, so there are real numbers in $[0, 1]$ that are not in S.

In fact, at least 9/10 of the interval is not in S. Moreover, we could rerun the argument with 1/100 (or 1/1000, 1/10000, and so on) in place of 1/10 to conclude that the fraction of $[0, 1]$ not in S is arbitrarily close to 1.   □

When the argument above is applied to the countable set of rational numbers between 0 and 1,

$$S = \left\{ \frac{1}{2}; \ \frac{1}{3}, \frac{2}{3}; \ \frac{1}{4}, \frac{3}{4}; \ \frac{1}{5}, \frac{2}{5}, \frac{3}{5}, \frac{4}{5}; \ \frac{1}{6}, \frac{5}{6}; \ \frac{1}{7}, \frac{2}{7}, \frac{3}{7}, \frac{4}{7}, \frac{5}{7}, \frac{6}{7}; \ \cdots \right\},$$

we conclude that irrational numbers exist in $[0, 1]$. Of course, we already knew this, but it is surprising that any number can escape being covered by any of the intervals U_n. After all, the rational numbers lie densely on the line, so intuition may suggest that covering each rational with an interval will cover all of $[0, 1]$.

Summing the infinite series $\frac{1}{10} + \frac{1}{10^2} + \frac{1}{10^3} + \cdots$ seems to refute this naive intuition, and there are two ways to obtain a clearer view.

1. If the intervals $U_1, U_2, U_3, \ldots$ cover all of $[0, 1]$ (somehow, despite their small total length), then we can show that *finitely many* of these intervals, say $U_1, U_2, \ldots, U_m$, also cover $[0, 1]$. A theorem to this effect is proved in the next section. Since the total length of $U_1, U_2, \ldots, U_m$ is less than 1/9, this is clearly absurd.

2. Since U_1 has length 1/10, one of the decimal fractions

$$0.0, \quad 0.1, \quad 0.2, \quad \ldots, \quad 0.9,$$

does not lie in U_1. Choose one of these fractions, say 0.2. Then, since U_2 has length 1/100, one of the decimal fractions

$$0.20, \quad 0.21, \quad 0.22, \quad \ldots, \quad 0.29$$

does not lie in U_2. Choose one of these fractions, and continue. At stage n we add an nth decimal place that keeps our number out of U_n. The infinite decimal thus obtained (with due precautions to avoid ambiguous decimals) therefore lies outside all of the intervals $U_1, U_2, U_3, \ldots$.

The second way of clarifying the measure argument is a strong hint at the diagonal argument. Indeed, the diagonal argument is precisely what comes to mind when one tries to find a specific number not covered by the intervals $U_1, U_2, U_3, \ldots$.

Exercises

Cantor first applied the uncountability of $\mathbb{R}$ to prove the existence of *transcendental* numbers; that is, nonalgebraic numbers.

3.5.1 Explain how the existence of transcendental numbers follows from the uncountability of $\mathbb{R}$.
3.5.2 Show, in fact, that "almost all" real numbers are transcendental.

The key idea of the diagonal argument—making a new object x that differs from the nth given object x_n in the nth place—works even better with subsets of $\mathbb{N}$ and sequences in $\mathbb{N}^{\mathbb{N}}$ than it does with real numbers (because the problem of "ambiguous objects" does not arise).

3.5.3 Given $S_1, S_2, S_3, \ldots \subseteq \mathbb{N}$, explain how to define an $S \subseteq \mathbb{N}$ such that $S \neq$ each S_n.
3.5.4 Given $f_1, f_2, f_3, \ldots \in \mathbb{N}^{\mathbb{N}}$, explain how to define an $f \in \mathbb{N}^{\mathbb{N}}$ such that $f \neq$ each f_n.

The uncountability of $\mathbb{R}$, and hence of $\mathcal{P}(\mathbb{N})$, leads to some surprising results about subsets of $\mathbb{N}$. Here are two such results, based on associating real numbers with sets or sequences of rationals, and exploiting the countability of $\mathbb{Q}$.

3.5.5 Show that there are uncountably many sets $S_x \subseteq \mathbb{N}$ such that, for any S_x, S_y, either $S_x \subseteq S_y$ or $S_y \subseteq S_x$.
3.5.6 Show that there are uncountably many sets $T_x \subseteq \mathbb{N}$, any two of which have only a finite intersection.

Cantor's first uncountability proof (in 1874) relied on the nested interval property from Sect. 2.6. Given countably many real numbers $x_1, x_2, x_3, \ldots$, he found an $x \neq x_1, x_2, x_3, \ldots$ in nested intervals defined as follows. Let $a_1 = x_1$,

$$b_1 = \text{first } x_i \text{ beyond } a_1 \text{ such that } x_i > a_1,$$

$$a_2 = \text{first } x_j \text{ beyond } b_1 \text{ such that } a_1 < x_j < b_1,$$

$$b_2 = \text{first } x_k \text{ beyond } a_2 \text{ such that } a_1 < a_2 < x_k < b_1,$$

and so on. Thus, $[a_1, b_1] \supset [a_2, b_2] \supset \cdots$.

3.5.7 If the sequence of intervals is finite, conclude that there is an $x \neq x_1, x_2, \ldots$.

3.5.8 If the sequence of nested intervals is infinite, conclude that any of its common points $x \neq$ $x_1, x_2, \ldots$.

3.6 Two Classical Theorems About Infinite Sets

The concept of a limit point, introduced in Sect. 2.6 in the case of infinite sequences, has an important generalization to infinite sets.

Definition. A point x is a *limit point* of a set S if, for any $\varepsilon > 0$, there is a point of S other than x within distance ε of x.

Limit points play an important role in mathematics, particularly in the two theorems below, which are crucial to later developments. These theorems illustrate how the concept of limit point is intimately related to the concept of infinite set, which we now know includes uncountable sets. For the sake of definiteness, we state the theorems as properties of the interval $[0,1]$, but they apply to any closed interval.

Bolzano–Weierstrass Theorem. *Any infinite set S of points in $[0, 1]$ has a limit point in $[0, 1]$.*

Proof. Since $I = [0, 1]$ contains infinitely many points of S, so does (at least) one half of I, either $[0, 1/2]$ or $[1/2, 1]$. To be specific, let

$$I_1 = \text{leftmost half of } I \text{ that contains infinitely many points of } S.$$

Similarly, let

$$I_2 = \text{leftmost half of } I_1 \text{ that contains infinitely many points of } S,$$

$$I_3 = \text{leftmost half of } I_2 \text{ that contains infinitely many points of } S,$$

and so on. Then $I, I_1, I_2, \ldots$ is a nested sequence of closed intervals, with lengths that become arbitrarily small, and hence with a common point x by the nested interval property of Sect. 2.6.

The point x is a limit point of S because in each I_n there are points of S other than x, and hence such points are arbitrarily close to x. $\square$

The next theorem concerns *open* intervals (a, b), exploiting the property that (a, b) contains, along with any member x, any sufficiently small interval containing x.

A set of open intervals may be uncountable, so when we denote a member of the set by U_i we allow the index i to range over a possibly uncountable set. We say that intervals U_i *cover* $[0, 1]$ if $[0, 1]$ is contained in the union of the U_i.

Heine-Borel Theorem. *If* $[0, 1]$ *is covered by infinitely many open intervals* U_i, *then* $[0, 1]$ *is covered by finitely many of the* U_i.

Proof. Suppose on the contrary that $I = [0, 1]$ can be covered only by infinitely many of the intervals U_i. Then it follows that some half of I, either $[0, 1/2]$ or $[1/2, 1]$, also can be covered only by infinitely many of the U_i. As in the previous proof, we make specific choice:

$$I_1 = \text{leftmost half of } I \text{ that can be covered only by infinitely many } U_i.$$

And similarly:

$$I_2 = \text{leftmost half of } I_1 \text{ that can be covered only by infinitely many } U_i,$$

$$I_3 = \text{leftmost half of } I_2 \text{ that can be covered only by infinitely many } U_i,$$

and so on. In this way we obtain a nested sequence of closed intervals $I, I_1, I_2, \ldots$, none of which can be covered by finitely many of the U_i. Since $I, I_1, I_2, \ldots$ become arbitrarily small, there is one point x common to them all, as in the previous proof.

But x belongs to some U_i (since the U_i cover all points of $[0, 1]$); call it U_j. Since U_j is open, it covers any sufficiently small I_k along with x. This contradicts the assumption that each I_k can *not* be covered by finitely many of the U_i.

Therefore, our original assumption, that $[0, 1]$ cannot be covered by finitely many of the U_i, is false.   □

The properties of $[0, 1]$ proved in the two theorems above reflect what is called its *compactness*. This property does not hold for the open interval $(0, 1)$, or for the whole line $\mathbb{R}$.

Definition. A set K is called *compact* if any cover of K by open intervals has a finite subcover.

The Heine–Borel theorem also enables us to clear up the problem raised in the second proof of uncountability in the previous section: whether the interval $[0, 1]$ can be covered by open intervals of total length < 1. If we have such a set of open intervals U_i covering $[0, 1]$, then finitely many of the U_i cover $[0, 1]$, and their total length is also < 1. Then, if we merge any overlapping members of this finite set into single open intervals, we obtain a *finite* set $\{V_1, V_2, \ldots, V_m\}$ of *disjoint* open intervals covering $[0, 1]$, with total length < 1. This is clearly impossible.

It was precisely to clear up this point that Borel (1895) introduced what we now call the Heine–Borel theorem. On page 51 of that paper he commented as follows

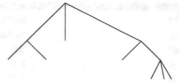

Fig. 3.7 Example of a tree

on the lemma that a closed interval I cannot be covered by open intervals with a total length less than that of I:

> One may regard this lemma as obvious; nonetheless, because of its importance, I wish to give a proof based on a theorem of interest in itself .... *If on a line [interval] there are an infinity of [open] intervals, so that each point of the line lies in at least one of the intervals, then one can effectively determine a* FINITE NUMBER *of intervals among the given intervals with the same property (that each point of the line is in the interior of at least one of them).*

Exercises

3.6.1 Give examples showing that the Bolzano–Weierstrass and Heine–Borel theorems do not hold with $(0,1)$ or $\mathbb{R}$ in place of $[0,1]$.

3.6.2 Use the Bolzano–Weierstrass theorem to show that nested interval property implies the least upper bound property, as claimed at the beginning of Sect. 2.6.

3.6.3 Prove that an infinite sequence of real numbers $x_1, x_2, x_3, \ldots$ (say, in $[0,1]$) contains either an infinite subsequence $y_1 < y_2 < y_3 < \cdots$ or an infinite subsequence $z_1 > z_2 > z_3 > \cdots$. (*Hint*: Look at a limit point of $x_1, x_2, x_3, \ldots$.)

The proofs of the Bolzano–Weierstrass and Heine–Borel theorems (and also Exercise 3.6.3) are based on the so-called "infinite pigeonhole principle." This principle says that, if an infinite set is divided into finitely many parts, then one of the parts is infinite.

Another theorem that begs to be proved by the infinite pigeonhole principle is the *Kőnig infinity lemma*, which states that an infinite tree whose vertices have finite degree has an infinite branch.

A tree is a structure like that shown in Fig. 3.7, in which there is a top vertex, connected to other vertices by edges, which are connected to other vertices in turn, in such a way that any two vertices are connected by a unique sequence of edges. The *degree* of any vertex is the number of edges connecting it to other vertices.

3.6.4 Prove that an infinite tree whose vertices have finite degree has an infinite branch, that is, an infinite sequence of vertices each connected by an edge to the one before.

3.6.5 Reinterpret the proof of the Bolzano–Weierstrass theorem as the construction of an infinite tree, whose infinite branches correspond to limit points.

3.7   The Cantor Set

A surprising and important uncountable set is one known as the *Cantor set*. This set, which we will call C for short, is also known as the "middle third" set because

Fig. 3.8 Early stages in the construction of the Cantor set

Fig. 3.9 Constructing the Cantor set via a tree

of the process that constructs it—removal of an infinite sequence of open intervals from the unit interval [0,1].

The first stage removes the middle third, (1/3,2/3), leaving the two closed intervals [0,1/3] and [2/3,1]. The second stage removes their middle thirds, leaving the four closed intervals [0,1/9], [2/9,1/3], [2/3,7/9], and [8/9,1]. The third stage removes their middle thirds, and so on. The results of the first six stages are shown in Fig. 3.8.

Each stage produces a finite union of closed intervals, each of which is 1/3 of an interval produced at the previous stage. Each nested sequence of intervals from successive stages produces exactly one point in C, because the lengths of the intervals tend to 0. Conversely, each point of C arises in this way, so the points of C correspond to the infinite paths down the *tree* shown in Fig. 3.9.

Each infinite path down the tree passes through a sequence of vertices, at each of which there is a choice to go left or right, corresponding to the choice of the left or right third of an interval. Thus, the points of C can be described by infinite sequences of the letters L and R. For example, the sequence $LLLL\ldots$ gives the point 0, and $RRRR\ldots$ gives the point 1.

An obvious diagonal argument shows that there are uncountably many infinite sequences of Ls and Rs, so C is an uncountable set. Indeed, it is clear that C is equinumerous with the set of infinite sequences of 0s and 1s, which was shown in Sect. 3.3 to be equinumerous with $\mathbb{R}$. Thus, C is equinumerous with $\mathbb{R}$, which is surprising, since C has measure zero!

3.7.1 Measure of the Cantor Set

We can find the measure of C by adding up the lengths of the intervals removed from $[0,1]$ in the construction of C.

In stage 1, the length removed $= 1/3$;

in stage 2, the length removed $= 2 \times 1/9 = 2/9 = 2/3^2$;

in stage 3, the length removed $= 4 \times 1/27 = 4/27 = 2^2/3^3$;

in stage 4, the length removed $= 8 \times 1/81 = 8/81 = 2^3/3^4$;

and in general

in stage $n + 1$, the length removed $= 2^n/3^{n+1}$.

So,

$$\text{total length removed} = \frac{1}{3}\left[1 + \frac{2}{3} + \left(\frac{2}{3}\right)^2 + \left(\frac{2}{3}\right)^3 + \cdots\right].$$

This is an instance of the general geometric series $a + ar + ar^2 + \cdots$ (with $a = 1/3$ and $r = 2/3$), which has sum $\frac{a}{1-r}$. Therefore, the total length of the intervals removed in the construction of C is

$$\frac{1/3}{1 - 2/3} = \frac{1/3}{1/3} = 1,$$

and hence the measure of C itself is zero.

Exercises

The removal process that creates C has a nice interpretation in terms of base 3 ("ternary") expansions of real numbers in $[0,1]$.

3.7.1 Explain why removing the middle third of $[0,1]$ leaves the numbers whose first ternary digit is 0 or 2.

3.7.2 Explain why removing the middle thirds of $[0,1/3]$ and $[2/3,1]$ leaves the numbers whose first and second digits are 0 or 2.

3.7.3 By continuing this argument, show that the numbers in C are those with ternary expansions that can be written entirely with 0s and 2s.

3.7.4 Use this ternary representation to give another proof that C is uncountable.

The *Sierpiński carpet* is a two-dimensional variant of the Cantor set, obtained by successively removing "middle thirds" from squares. Figure 3.10 shows the first three approximations to the Sierpinski carpet.

3.7.5 Show that the area of the Sierpinski carpet is zero.

Fig. 3.10 First three approximations to the Sierpinski carpet

3.8 Higher Cardinalities

So far we have seen infinite sets of two different cardinalities: those with the cardinality of $\mathbb{N}$ and those with the cardinality of $\mathbb{R}$. Moreover, $\mathbb{R}$ is of *higher cardinality* than $\mathbb{N}$ in the sense that there is an injection from $\mathbb{N}$ into $\mathbb{R}$, but no bijection, because of the diagonal argument. Cantor (1891) noticed that the diagonal argument may be applied to any set X to produce a set of higher cardinality than X; namely, the *power set* $\mathcal{P}(X)$ whose members are the subsets of X. So in fact there are infinite sets of infinitely many different cardinalities.

Cantor's Theorem on the Power Set. *For any set X, there are more subsets of X than there are elements.*

Proof. Consider any pairing $x_i \leftrightarrow X_i$ between the elements x_i of X and certain subsets X_i of X. No matter how the pairing is made, the sets X_i do not include all the subsets of X because they do not include the *diagonal set D* defined by the property

$$x_i \in D \Leftrightarrow x_i \notin X_i.$$

Indeed D differs from each X_i with respect to the element x_i; if x_i is X_i then x_i is *not* in D, and if x_i is not in X_i then x_i *is* in D.

Thus, there are more subsets of X than there are elements of X. In other words, the set $\mathcal{P}(X)$ has higher cardinality than X.   □

It follows in particular that subsets of $\mathbb{R}$ are more numerous than the real numbers. It happens that the subsets of $\mathbb{R}$ that come most naturally to mind (the Borel sets, see Chap. 8) form a collection with only as many members as $\mathbb{R}$. Thus, we have the opportunity to use the diagonal argument to find new sets of real numbers beyond the obvious ones—much as we used the diagonal argument, in Exercise 3.5.1, to find new real numbers beyond the algebraic numbers.

3.8.1   The Continuum Hypothesis

With the discovery that there are different kinds of infinity, two questions arise:

1. Does $\mathbb{R}$ represent the smallest uncountable infinity? In particular, is there an uncountable set of real numbers not equinumerous with $\mathbb{R}$?
2. Is the diagonal method essentially the only way to prove the existence of uncountable sets?

The conjecture that any uncountable set of real numbers is equinumerous with $\mathbb{R}$ was first posed by Cantor (1878), and it is the first version of what is called the *continuum hypothesis*. We will discuss this hypothesis, which is not yet settled, further in Chap. 5. There we will also show that the diagonal method is *not* the only way to prove the existence of uncountable sets. There is another method, also discovered by Cantor, involving the so-called *ordinal numbers*. The concept of ordinal number also leads to a sharper statement of the continuum hypothesis, and to the clarification of *axioms* for set theory, which will be the subject of Chap. 6.

3.8.2   Extremely High Cardinalities

It will become clear when we discuss axioms for set theory in Chap. 6 that iteration of the power set operation $\mathcal{P}$ can produce sets of extraordinarily high cardinality. Just to give a taste of what is possible, consider the sequence

$$\mathbb{N}, \quad \mathcal{P}(\mathbb{N}), \quad \mathcal{P}(\mathcal{P}(\mathbb{N})), \quad \ldots.$$

Each set in the sequence has cardinality greater than the one before, so the union of all these sets,

$$Y = \mathbb{N} \cup \mathcal{P}(\mathbb{N}) \cup \mathcal{P}(\mathcal{P}(\mathbb{N})) \cup \cdots,$$

has cardinality greater than any of $\mathbb{N}, \mathcal{P}(\mathbb{N}), \mathcal{P}(\mathcal{P}(\mathbb{N})), \ldots$. Then, of course, $\mathcal{P}(Y)$ has cardinality greater than Y, and so on. One wants to say "ad infinitum," but it is no longer clear what that means—infinity is certainly bigger than we first thought.

Despite the immense power of set theory to produce sets of high cardinality, there are "largeness" properties so exorbitant that sets with such properties cannot be proved to exist. For example, a set Z is called *inaccessible* if

1. Z has infinite members,
2. $X \in Z$ implies $\mathcal{P}(X) \in Z$, and
3. $X \in Z$ implies that the range of any function with domain X and values in Z is a member of Z.

The existence of inaccessible sets is not provable from the standard axioms of set theory, for reasons that will emerge in Chap. 6. Here is a clue (though it will probably not help at this point): if an inaccessible set exists, then its existence is not provable!

Exercises

3.8.1 Show that $\mathcal{P}(\mathcal{P}(\mathbb{N}))$ is equinumerous with the set of real functions.

The power set operation is also interesting when applied to finite sets, starting with the empty set (denoted by { } or $\emptyset$).

3.8.2 If $F = \{x_1, \ldots, x_n\}$ is a set with n elements, show that $\mathcal{P}(F)$ has 2^n elements.

3.8.3 According to Exercise 3.8.2, $\mathcal{P}(\emptyset)$ has one element and $\mathcal{P}(\mathcal{P}(\emptyset))$ has two. Write down these elements.

Iterating the power set operation $\mathcal{P}$ any finite number of times, starting with the empty set, gives an important series of sets V_n, defined inductively as follows.

$$V_0 = \emptyset,$$

$$V_{n+1} = V_n \cup \mathcal{P}(V_n).$$

3.8.4 Prove that any subset of V_n is a member of V_{n+1}, and that if $X \in V_n$ then $\mathcal{P}(X) \in V_{n+1}$.

We now define V_ω to the union of all the V_n.

3.8.5 Show that V_ω satisfies the last two conditions for inaccessibility, but not the first.

3.8.6 Give an example of a set that satisfies the first two conditions for inaccessibility, but not the last.

3.9 Historical Remarks

When Cantor discovered, in 1874, that infinite sets can be countable or uncountable, almost all previous thinking about infinity was superseded. For example, the vague and contentious distinction between "potential" and "actual" infinity was replaced by the clear and dramatic distinction between countable and uncountable. It seems academic to debate whether the positive integers $1, 2, 3, \ldots$ should be viewed as a collection that grows, one member at a time, or as a completed whole $\mathbb{N} = \{1, 2, 3, \ldots\}$, once it is known that $\mathbb{R}$ *cannot* be viewed as a collection that grows one member at a time, because it is uncountable.

The discovery of uncountability also brought new clarity to the concept of countability. Before 1874, few examples of countably infinite sets were actually known, apart from $\mathbb{N}$ and some of its subsets. It is thought that Cantor noticed the countability of $\mathbb{Q}$ some time before he proved the uncountability of $\mathbb{R}$. In the intervening period, Cantor asked Dedekind whether he could prove that $\mathbb{R}$ is

Fig. 3.11 Georg Cantor

countable. Dedekind was unable to do so, but he offered a proof that the algebraic numbers are countable (example 6 of Sect. 3.1). Ironically, Dedekind's result became the centerpiece of Cantor's paper on the uncountability of $\mathbb{R}$.

Just how this came about is explained by Ferreirós (1999), pp. 175–180. Apparently, Weierstrass persuaded Cantor to play down his uncountability proof in favor of its more topical corollary: the existence of transcendental numbers. Such numbers were first exhibited by Liouville (1851) and, just one year before Cantor's discovery, Hermite (1873) had proved that e is transcendental. Compared with these, Cantor's transcendence proof was remarkably simple—and to this day it is the most elementary proof known. This is why Cantor (1874) bears the (to us) unenlightening title "On a property of the collection of all real algebraic numbers."

Another theorem for which Dedekind deserves the credit is the so-called Cantor–Schröder–Bernstein theorem. Before any of these three had a proof, Dedekind found one in 1887, but omitted it (except for a key lemma) from the book, Dedekind (1888), he was then writing (see Ferreirós 1999, Chap. VII). For a long time, Dedekind had been interested in mappings between infinite sets, and in 1882 he proposed to *define* an infinite set as one that admits a bijection with a proper subset of itself (see Dedekind 1888, Sect. 64). It is indeed clear that a set admitting such a bijection must be infinite, but it is a more subtle question whether every infinite set admits such a bijection. We take up this issue in Sect. 7.1.

The Bolzano–Weierstrass theorem of Sect. 3.6 gets its name because Bolzano (1817) used the bisection argument, and a sequence of nested intervals, in his attempt to prove the intermediate value theorem for continuous functions. As

Fig. 3.12 Henry John Stephen Smith

we mentioned in Sect. 1.6, Bolzano's proof lacked a definition of $\mathbb{R}$ that could justify any assumption of completeness, such as the nested interval principle. Weierstrass (1874) revisited the theorem after definitions of $\mathbb{R}$ had been proposed— by Dedekind, Cantor, and himself—at which time the nested interval argument was justifiable.

The related Heine–Borel theorem likewise began with an argument, due to Heine (1872), aimed at a different theorem: in this case the theorem that a continuous function on a closed interval is uniformly continuous. (See Sects. 4.6 and 4.7 for this theorem and a discussion of uniform continuity.) Borel (1895) was the first to prove the theorem in its present form, and also the first to recognize its importance in measure theory. In particular, Borel saw that the Heine–Borel theorem justifies the measure argument that $\mathbb{R}$ is uncountable.

Harnack (1885) had observed, as we did in Sect. 3.5, that any countable set could be covered by intervals of arbitrarily small total measure. But Harnack was puzzled by the example of the countable set $\mathbb{Q}$. Thinking that a covering of $\mathbb{Q}$ by intervals would cover *all* points, he jumped to the conclusion that the whole interval [0,1] could be covered by open intervals of total length ε. This of course plays havoc with the concept of measure, and fortunately the Heine–Borel theorem showed that Harnack was wrong. For this reason, Borel (1895) called the Heine–Borel theorem the "first fundamental theorem of measure theory." (For more details on Harnack's mistake, see Bressoud 2008, p. 63.)

The so-called Cantor set is actually due to Smith (1875), shown in Fig. 3.12. However, the set plays so many important roles—in the theory of $\mathbb{R}$, continuous functions, and measure theory—that credit for it can be shared among several

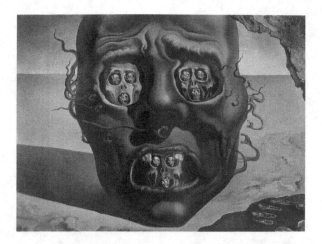

Fig. 3.13   Dali's *Face of War*

mathematicians. There are also higher-dimensional variations, such as the Sierpiński carpet, and even artists have hit upon a similar idea.

Figure 3.13 shows a Cantor-style set in the work of Salvador Dali.

The Cantor (1874) argument for the uncountability of $\mathbb{R}$ constructs a real number x unequal to each member of a given sequence $x_1, x_2, x_3, \ldots$. But it is not a "diagonal" argument in the sense of making x unequal to x_n at a predetermined decimal place. An argument closer to diagonalization occurs in du Bois-Reymond (1875), but Cantor does not seem to have been influenced by it. In any case, the diagonal argument in Cantor (1891) is clearly simpler and more general. Even Cantor may have been surprised that it was so easy to prove that there was no largest set—a result that he had conjectured before but only on vague grounds.

But with the proof came other concerns. If there is no largest set, there is no set of all sets. This may be a problem—but it also may be a useful clarification. Certainly, one can no longer suppose that, for each property P, there is the set of all objects with property P (let P be the property of being a set). For some, this was a "crisis of foundations"; for others (who ultimately prevailed) it was an argument for the *cumulative* or *hierarchical* concept of set. In the cumulative concept, all sets arise from the empty set $\emptyset$ by certain operations, just as natural numbers arise from 0 by the successor operation. From the cumulative viewpoint, it makes no more sense to have the "set of all sets" than it does to have the "number of all numbers."

In Chap. 6 we will see exactly how the cumulative concept of set unfolds, and in Chaps. 6 and 9 we will touch on the question of sets so large that their existence is not provable.

Chapter 4
Functions and Limits

PREVIEW

Now that we are familiar with the real numbers, we can better understand some basic concepts of analysis: limits, convergence, and continuity. In particular, the absence of gaps in $\mathbb{R}$ explains the absence of gaps in the graph of any continuous function. This, and other "obvious" properties of continuous functions, depends directly on the completeness of $\mathbb{R}$.

On the other hand, continuous functions sometimes have extremely surprising properties. We construct three examples:

- A continuous function that increases from 0 to 1 while remaining "constant almost everywhere."
- A curve with no tangents.
- A curve that fills a square.

The last example raises the question: is there a continuous bijection between the line interval [0,1] and the square $[0, 1] \times [0, 1]$? We show that the answer is no, as one would hope if the concept of dimension is to be meaningful. The key to the proof is the so-called *intermediate value theorem* about the absence of gaps in the graph of a continuous function.

Finally, we explain why continuity ensures that a function has an integral. In fact, to integrate a continuous function one needs only the simplest concept of integral: the *Riemann integral* familiar from basic calculus.

4.1 Convergence of Sequences and Series

In Sect. 2.6 we touched on the concepts of *convergence* and *limit* for sequences; namely, a sequence $c_1, c_2, c_3, \ldots$ has limit c if, for each number $\varepsilon > 0$, there is a natural number N such that

$$n > N \Rightarrow |c_n - c| < \varepsilon.$$

J. Stillwell, *The Real Numbers: An Introduction to Set Theory and Analysis*,
Undergraduate Texts in Mathematics, DOI 10.1007/978-3-319-01577-4_4,
© Springer International Publishing Switzerland 2013

This relation is written $\lim_{n\to\infty} c_n = c$ for short. One of the most important uses of the limit concept for sequences is in defining the sum of an infinite series:

Definition. An infinite series $a_1 + a_2 + a_3 + \cdots$ is said to *converge* to *sum s* if the sequence of partial sums

$$a_1, \quad a_1 + a_2, \quad a_1 + a_2 + a_3, \quad \cdots$$

has limit s.

A familiar example of a convergent infinite series is the geometric series

$$1 + a + a^2 + a^3 + \cdots \quad \text{for} \quad |a| < 1.$$

This series converges because the nth partial sum is

$$s_n = 1 + a + a^2 + \cdots + a^{n-1},$$

for which we find

$$as_n = a + a^2 + a^3 + \cdots + a^{n-1} + a^n.$$

Subtraction then gives $s_n = \frac{1-a^n}{1-a}$, which has limit $\frac{1}{1-a}$ when $|a| < 1$.

Infinite decimals can also be viewed as infinite series, comparable with geometric series. For example

$$\sqrt{2} = 1.4142135 \cdots$$

$$= 1 + \frac{4}{10} + \frac{1}{10^2} + \frac{4}{10^3} + \frac{2}{10^4} + \frac{1}{10^5} + \frac{3}{10^6} + \frac{5}{10^7} + \cdots$$

Since all terms of this series are ≥ 0, the partial sums s_n increase with n. Also, they are bounded above by the geometric series

$$9 + \frac{9}{10} + \frac{9}{10^2} + \frac{9}{10^3} + \frac{9}{10^4} + \cdots = \frac{9}{1 - 1/10} = 10.$$

Thus, the series converges, to the lub of the set of its partial sums.

4.1.1 Divergent and Conditionally Convergent Series

For a series $a_1 + a_2 + a_3 + \cdots$ to converge it is necessary that the nth term a_n have limit zero, otherwise the partial sums will not become arbitrarily close to any limit value. However, it is not sufficient for a_n to have limit zero, as is shown by the famous example of the *harmonic series*:

$$\frac{1}{2} + \frac{1}{3} + \frac{1}{4} + \frac{1}{5} + \cdots .$$

This series is said to *diverge*, because its partial sums grow in size indefinitely, as one can see by grouping terms as follows:

$$\frac{1}{2} + \left(\frac{1}{3} + \frac{1}{4}\right) + \left(\frac{1}{5} + \frac{1}{6} + \frac{1}{7} + \frac{1}{8}\right) + \left(\frac{1}{9} + \frac{1}{10} + \cdots + \frac{1}{16}\right) + \cdots .$$

Each group has sum at least 1/2, so by taking enough groups we can make the partial sum as large as we please.

Similar arguments lead to the conclusion that the series

$$1 + \frac{1}{3} + \frac{1}{5} + \frac{1}{7} + \cdots \quad \text{and} \quad \frac{1}{2} + \frac{1}{4} + \frac{1}{6} + \frac{1}{8} + \cdots$$

both diverge. This leads to an interesting property of the series

$$1 - \frac{1}{2} + \frac{1}{3} - \frac{1}{4} + \frac{1}{5} - \frac{1}{6} + \frac{1}{7} - \frac{1}{8} + \cdots ,$$

called *conditional convergence*.

The partial sums obtained by taking terms in the order shown above, namely,

$$1, \quad 1 - \frac{1}{2}, \quad 1 - \frac{1}{2} + \frac{1}{3}, \quad 1 - \frac{1}{2} + \frac{1}{3} - \frac{1}{4}, \quad \cdots$$

fall inside a sequence of nested intervals

$$\left[1 - \frac{1}{2}, 1\right], \quad \left[1 - \frac{1}{2}, 1 - \frac{1}{2} + \frac{1}{3}\right], \quad \left[1 - \frac{1}{2} + \frac{1}{3} - \frac{1}{4}, 1 - \frac{1}{2} + \frac{1}{3}\right], \quad \cdots$$

the length of which becomes arbitrarily small (because the nth interval has length $1/(n + 1)$). The partial sums therefore have a limit by the nested interval property of Sect. 2.6, *when summed in the above order*.[1]

But if we allow the terms to be rearranged, then the series can converge to any number we please! This comes about because we can collect enough of the positive terms

$$1, \quad \frac{1}{3}, \quad \frac{1}{5}, \quad \frac{1}{7}, \quad \cdots$$

---

[1] In fact, the sum is log 2, the natural logarithm of 2, as you may recall from calculus.

to exceed any number we please, and the same is true of the negative terms

$$-\frac{1}{2}, \quad -\frac{1}{4}, \quad -\frac{1}{6}, \quad -\frac{1}{8}, \quad \cdots$$

Thus, if we want the series to have the sum 3/2, for example, we do the following:

- Collect terms $1, \frac{1}{3}, \frac{1}{5}, \ldots$ until their sum exceeds 3/2. This happens when we get to $1 + \frac{1}{3} + \frac{1}{5}$.
- Then add negative terms until the sum falls below 3/2. This happens as soon as $-\frac{1}{2}$ is added.
- Resume adding positive terms until the sum exceeds 3/2 again. This happens with $1 + \frac{1}{3} + \frac{1}{5} - \frac{1}{2} + \frac{1}{7} + \cdots + \frac{1}{15}$.
- Then resume adding negative terms (in this case just $-\frac{1}{4}$) until the sum falls below 3/2 again, and so on.

The partial sums thereby oscillate on either side of 3/2, but they approach it ever more closely, since the terms of the series become ever closer to 0. Therefore, the sum of the series is precisely 3/2. By a similar argument we can arrange terms so that the sum is any number we please.

Exercises

The proof that the harmonic series diverges relies on collecting groups of terms that sum to at least 1/2. Each group of terms contains twice as many terms as the preceding group, so to exceed the sum k one needs around 2^k terms. This estimate suggests that the sum of the first n terms is around the size of $\log n$. We can show that this estimate is remarkably accurate by comparing the sum $1 + \frac{1}{2} + \frac{1}{3} + \cdots + \frac{1}{n}$ with the geometric interpretation of $\log n$ as an area under the curve $y = 1/x$.

We compare the two as shown in Fig. 4.1. The natural logarithm $\log n$ is the area under $y = 1/x$ from 1 to n, while $1 + \frac{1}{2} + \frac{1}{3} + \cdots + \frac{1}{n}$ is the total area of a collection of rectangles. Each rectangle has width 1, and their heights are $1, \frac{1}{2}, \frac{1}{3}, \ldots, \frac{1}{n}$ respectively.

4.1.1 By referring to Fig. 4.1, explain why $\log n < 1 + \frac{1}{2} + \frac{1}{3} + \cdots + \frac{1}{n}$.

4.1.2 With the help of a similar figure, explain why $\log n > \frac{1}{2} + \frac{1}{3} + \cdots + \frac{1}{n}$.

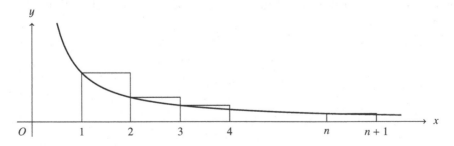

Fig. 4.1 Comparing the logarithm with the harmonic series

4.1.3 Deduce from the figures in Exercises 4.1.1 and 4.1.2 that the terms

$$c_n = 1 + \frac{1}{2} + \frac{1}{3} + \cdots + \frac{1}{n} - \log n$$

form a bounded increasing sequence, with limit ≤ 1.

The limit of the sequence $c_1, c_2, c_3, \ldots$ is known as *Euler's constant* γ, and it is approximately 0.57721. Although γ has been much studied—Havil (2003) is a whole book about it—we do not yet know whether γ is irrational.

The ability of a conditionally convergent series to represent any real number gives an interesting proof that $\mathbb{R}$ *is equinumerous with the set of permutations of* $\mathbb{N}$. (A *permutation* of $\mathbb{N}$ is bijection: $\mathbb{N} \to \mathbb{N}$. Informally, a permutation is a "rearrangement.")

4.1.4 Use a conditionally convergent series to define an injection from $\mathbb{R}$ into {permutations of $\mathbb{N}$}.

4.1.5 Using binary sequences, say, define an injection from {permutations of $\mathbb{N}$} into $\mathbb{R}$.

4.1.6 Hence prove that the set {permutations of $\mathbb{N}$} is equinumerous with $\mathbb{R}$.

4.2   Limits and Continuity

When we try to capture the notion of *continuity*—for example, to say what it means for a graph or a curve to be "unbroken"—we know that we have to fall back on the completeness of $\mathbb{R}$. Completeness is related to limits, by Sect. 2.6, so it is no surprise that continuity involves the limit concept. The limit of a real function is defined as follows.

Definition. A real function f is said to have *limit l as x tends to a* if, for each $\varepsilon > 0$, there is a $\delta > 0$ such that

$$0 < |x - a| < \delta \Rightarrow |f(x) - l| < \varepsilon.$$

This relationship can be expressed informally by saying that "$f(x)$ approaches l as x approaches a," and the notation for it is $\lim_{x \to a} f(x) = l$. The reason for the condition $0 < |x - a|$ is that we really want to express the behavior as x *approaches* a—not what happens when $x = a$. For example, the function

$$f(x) = \begin{cases} 0 \text{ for } x \neq 0 \\ 1 \text{ for } x = 0. \end{cases}$$

has limit 0 as x approaches 0, even though $f(0) \neq 0$. We would not want to say, however, that this function is continuous at 0, since its graph has a clear break there. The definition of continuity says that the limit exists *and* that it equals the function value:

Definition. A real function f is *continuous at x = a* if $\lim_{x \to a} f(x) = f(a)$, and f is *continuous on a set S* (typically $S = \mathbb{R}$ or some interval) if f is continuous at each point of S.

These definitions seem to capture our intuitive concept of continuity, inasmuch as the continuous functions include the functions that obviously have "unbroken" graphs, and they exclude functions with graphs that are obviously "broken" in some way. Here are some examples.

1. Constant functions $f(x) = c$ are continuous, as is the identity function $f(x) = x$. For a constant function $f(x) = c$ we have

$$|x - a| < \delta \Rightarrow |f(x) - f(a)| < \varepsilon$$

for any δ whatever, because in fact $|f(x) - f(a)| = |c - c| = 0$.
   For the identity function $f(x) = x$ it suffices to choose $\delta = \varepsilon$, because then

$$|x - a| < \delta \Rightarrow |x - a| < \varepsilon \Rightarrow |f(x) - f(a)| < \varepsilon,$$

since $f(x) = x$ and hence $f(a) = a$.

2. If f_1 and f_2 are continuous functions, then so are $f_1 + f_2$, $f_1 - f_2$, $f_1 \cdot f_2$ and (at points where $f_2 \neq 0$) f_1/f_2. Thus, it follows, from the previous example, that all rational functions are continuous at the points where they are defined.
   Here we will explain why $f_1 + f_2$ is continuous; the proofs for the other cases are indicated in the exercises. Suppose that f_1 and f_2 are both continuous at $x = a$. We want to prove that $f_1 + f_2$ is also continuous there. This means proving, given $\varepsilon > 0$, that there is a $\delta > 0$ such that

$$0 < |x - a| < \delta \Rightarrow |f_1(x) + f_2(x) - f_1(a) - f_2(a)| < \varepsilon.$$

Now the continuity of f_1 gives us a δ_1 such that

$$0 < |x - a| < \delta_1 \Rightarrow |f_1(x) - f_1(a)| < \varepsilon/2,$$

and the continuity of f_2 gives us a δ_2 such that

$$0 < |x - a| < \delta_2 \Rightarrow |f_2(x) - f_2(a)| < \varepsilon/2.$$

So if we take $\delta = \min(\delta_1, \delta_2)$ we have

$$0 < |x - a| < \delta \Rightarrow |f_1(x) - f_1(a)| < \varepsilon/2 \quad \text{and} \quad |f_2(x) - f_2(a)| < \varepsilon/2$$
$$\Rightarrow |f_1(x) + f_2(x) - f_1(a) - f_2(a)| < \varepsilon/2 + \varepsilon/2 = \varepsilon,$$

as required.

3. The function $f(x) = 1/x$ is *not* continuous at $x = 0$.
   No matter what value we give to $f(0)$, $f(x)$ has no limit at all as x approaches 0. When x approaches 0 from the right $1/x$ grows beyond all positive bounds, and from the left $1/x$ grows beyond all negative bounds.

4. The *Dirichlet function*

$$f(x) = \begin{cases} 1 \text{ if } x \text{ is rational} \\ 0 \text{ if } x \text{ is irrational.} \end{cases}$$

is not continuous at any point.

This is because $\lim_{x \to a} f(x)$ does not exist at any point $x = a$. Any interval on the line contains both rational and irrational points, so $|f(x) - f(a)|$ varies by 1 in any interval, and hence cannot be made smaller than every $\varepsilon > 0$.

It may seem that this function is as discontinuous as it can possibly be, but we will see in Chap. 8 that the Dirichlet function is, in a sense, not far removed from continuous functions.

In some sense, continuous functions are the simplest functions, and discontinuous functions can indeed be very complicated. However, continuous functions can also have interesting complications, as we will see in the next section.

To conclude this section we observe a consequence of continuity that involves limits of sequences rather than limits of functions.

Sequential Continuity. *If f is continuous at a and $a_1, a_2, a_3, \ldots$ is any sequence with limit a, then the sequence $f(a_1), f(a_2), f(a_3), \ldots$ has limit $f(a)$.*

Proof. Since f is continuous at $x = a$, for each $\varepsilon > 0$ there is a $\delta > 0$ such that

$$0 < |x - a| < \delta \Rightarrow |f(x) - f(a)| < \varepsilon.$$

Now, since $a_1, a_2, a_3, \ldots$ has limit a, for this $\delta > 0$ there is an N such that

$$n > N \Rightarrow |a_n - a| < \delta \Rightarrow |f(a_n) - f(a)| < \varepsilon,$$

and this means that the sequence $f(a_1), f(a_2), f(a_3), \ldots$ has limit $f(a)$. □

This result has, as a corollary, the property of continuous functions that we used in Sect. 3.4:

Corollary 1. *Any continuous function on $\mathbb{R}$ is determined by its values on the rational numbers.*

Proof. Any irrational number a is the limit of a sequence of rational numbers $a_1, a_2, a_3, \ldots$. But then $f(a)$ is determined by the values of f on the rational numbers, namely, $f(a) = \lim_{n \to \infty} f(a_n)$. □

Exercises

4.2.1 By an argument like that used above to prove that $f_1 + f_2$ is continuous at $x = a$ if f_1 and f_2 are, prove that $f_1 - f_2$ is also continuous at $x = a$.

Fig. 4.2 Graph of the Thomae function

4.2.2 Use the identity

$$f_1(x)f_2(x) - f_1(a)f_2(a) = f_1(x)[f_2(x) - f_2(a)] + f_2(a)[f_1(x) - f_1(a)]$$

to prove that, if f_1 and f_2 are continuous at $x = a$, then so is $f_1 f_2$.

4.2.3 Prove that, if f is continuous at $x = a$ and $f(a) \neq 0$, then $1/f$ is continuous at $x = a$.

4.2.4 Deduce from the previous exercises that if f_1 and f_2 are continuous at $x = a$, and if $f_2(a) \neq 0$, then f_1/f_2 is continuous at $x = a$.

4.2.5 Use sequential continuity to prove that $f(g(x))$ is continuous at $x = a$ if f and g are. (The question whether sequential continuity implies continuity will be taken up in Sect. 7.1.) Also prove this assuming only ordinary continuity.

Another interesting discontinuous function is the *Thomae function* (also known as the "popcorn function"), due to Thomae (1879) and defined by

$$t(x) = \begin{cases} 1/q & \text{if } x \text{ is rational and } x = p/q \text{ in lowest terms} \\ 0 & \text{if } x \text{ is irrational.} \end{cases}$$

Figure 4.2 shows an approximation to its graph.

Despite its similarity to the Dirichlet function, the Thomae function is not discontinuous everywhere.

4.2.6 Show that $t(x)$ discontinuous at $x = p/q$ but continuous elsewhere.

4.3   Two Properties of Continuous Functions

The suitability of our definition of continuous functions is confirmed by the following two theorems, which express properties of continuous functions that our intuition expects.

Intermediate Value Theorem. *If f is continuous on interval $[a, b]$, with $f(a) < 0$ and $f(b) > 0$, then there is a value $c \in [a, b]$ with $f(c) = 0$.*

Proof. Consider the set of points x for which f is negative "all the way up to x":

$$S = \{x \in [a,b] : f(y) < 0 \text{ for all } y \le x\}.$$

S is nonempty, because a is a member, and bounded above by b, so S has a least upper bound c. Now consider whether $f(c)$ can be nonzero.

If $f(c) = \varepsilon > 0$ then, by continuity, there is a $\delta > 0$ with $f(c - \delta) > 0$. Since $c - \delta \in S$, this contradicts the definition of S. If $f(c) = -\varepsilon < 0$ then there is similarly a $\delta > 0$ with $f(y) < 0$ for $y < c + \delta$, which again contradicts the definition of S.

Thus, the only possibility is $f(c) = 0$. $\qquad\square$

The first to realize that the intermediate value property was provable was Bolzano (1817), and he also realized that the least upper bound property was the key to the proof. However, the least upper bound property was not available until Dedekind gave a precise definition of real numbers, by Dedekind cuts, in 1858. Once the least upper bound property was established, all the basic theorems about continuous functions became provable. The next theorem is another example.

Extreme Value Theorem. *A continuous function on a closed interval takes a maximum value (and, similarly, a minimum value).*

Proof. To create an opportunity to use the least upper bound property, we first prove that a continuous function f has a bounded set of values on the closed interval $[a, b]$.

If not, repeatedly bisect the interval $[a, b]$, each time choosing the leftmost half in which f has arbitrarily large values. In this way we obtain a sequence of closed intervals $I_1, I_2, I_3, \ldots$, in each of which f is unbounded, yet the length of the intervals tends to zero. It follows that the intervals have a single common point, $c \in [a, b]$. Then, by continuity, in a sufficiently small interval I_n containing c, the value of f remains within distance ε of $f(c)$—so f is *not* unbounded on I_n.

This contradiction shows that the set $\{f(x) : x \in [a,b]\}$ is bounded, and so it has a least upper bound l. If f does not take the value l, then the function $1/(l - f(x))$ is continuous on $[a, b]$. But then $1/(l - f(x))$ is bounded, by the argument above, which contradicts the assumption that $l - f(x)$ becomes arbitrarily small.

The latter contradiction shows that $f(x)$ takes the value l, which is necessarily its maximum value. Similarly, the continuous function $-f(x)$ takes a maximum value m, in which case $-m$ is the minimum value of $f(x)$. $\qquad\square$

4.3.1 The Devil's Staircase

In this subsection we discuss a function that throws new light on the intermediate value theorem. It is a function that takes all values between 0 and 1, while "almost never" changing value. For this function, both continuity and the intermediate value property verge on the miraculous.

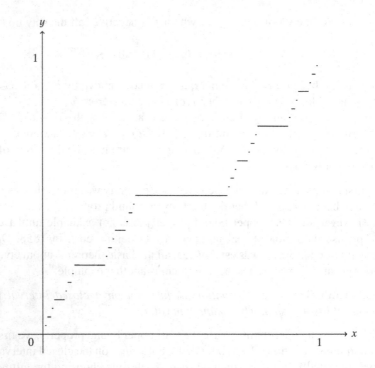

Fig. 4.3 Stage 5 in the construction of the Devil's staircase

The *Devil's staircase* is the graph of a continuous function F that is constant on each interval in the complement of the Cantor set. F is constructed in stages as follows:

Stage 1. Let $F(0) = 0$, $F(1) = 1$, and $F(x) = 1/2$ on the interval $(1/3, 2/3)$.

Stage 2. Let $F(x) = 1/4$ on the interval $(1/9, 2/9)$, and let $F(x) = 3/4$ on the interval $(7/9, 8/9)$.

Stage *n*. On the middle third of each interval (a, b) on which F is still undefined, but such that $F(a)$ and $F(b)$ are defined, let F take the value halfway between $F(a)$ and $F(b)$.

Figure 4.3 shows the graph of the function F at Stage 5.

After all finite stages are completed (so F is defined on all intervals in the complement of the Cantor set), any x for which $F(x)$ is still undefined will be arbitrarily close to points for which $F(x)$ is defined. Moreover, the difference between defined values $F(a), F(b)$ for $a < x < b$ becomes arbitrarily small as a and b approach x, so we can define $F(x)$ uniquely as the limit of the values $F(a)$ as $a \to x$.

It follows that F is a continuous function with values including all the binary fractions $m/2^n$ between 0 and 1, and all the limit points of these fractions—that is, all the real numbers between 0 and 1. In fact, the values of F *on the Cantor set itself* include all the binary fractions (as values at the endpoints of intervals in the complement) and their limit points, so F even maps the Cantor set continuously onto [0,1].

Exercises

A simple consequence of the intermediate value theorem is the following special case (the one-dimensional case) of the famous *Brouwer fixed-point theorem*: *Any continuous map $f : [0, 1] \rightarrow [0, 1]$ has a fixed point; that is, a value $c \in [0, 1]$ such that $f(c) = c$.*

4.3.1 Show that any continuous $f : [0, 1] \rightarrow [0, 1]$ has a fixed point, by considering intermediate values of a suitable function.

4.3.2 Give an example to show that a continuous function on an open interval need not have extreme values.

An explicit continuous map f of C onto [0, 1] may be described as follows. Recall from the exercises to Sect. 3.7 that each $x \in C$ has a unique ternary expansion using only the digits 0 and 2. Let f send this x to the number whose *binary* expansion has 1 in each place where the ternary expansion of x has a 2.

4.3.3 Explain why f is onto [0,1].

4.3.4 Show that $f(x)$ and $f(x')$ differ by less than 2^{-n} if x and x' differ by less than 3^{-n}, so that f is continuous.

4.4 Curves

We define a *curve* (strictly, a *curve with endpoints*, which may be identical), in the plane $\mathbb{R}^2$ say, to be a continuous function $f : [0, 1] \rightarrow \mathbb{R}^2$. This formalizes the idea of tracing the curve by moving a point along it in a unit time interval: $f(t)$ is the position of the point at time t. It might be thought sufficient to take the *range* of the function f to be the curve. An example that explains why this is not sufficient is the second one in this section: a continuous curve that fills a square.

Our first example is another that debunks a long-held belief about curves: that they always have tangents.

4.4.1 A Curve Without Tangents

Helge von Koch (1904) gave a lovely example of a curve without tangents, obtained as the limit of the sequence of polygons shown in Fig. 4.4.

We will not give a rigorous proof that the Koch curve has no tangents, but rather encourage the reader to visualize a process of repeated magnification of the curve. Unlike a smooth curve, which looks more and more like a straight

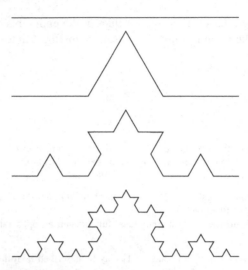

Fig. 4.4 The Koch polygon sequence

line under magnification, any portion of the Koch curve looks *exactly the same* when magnified by 3. If the Koch curve had tangents, magnification would show it becoming straighter in the neighborhood of any point where a tangent exists.

4.4.2   A Space-Filling Curve

Following Peano (1890), we prove that such a curve exists by describing how to move a point continuously through the square in a unit time interval, so that each point in the square is visited at some time between 0 and 1.

Peano's space-filling curve. *There is a continuous surjection f of the unit interval* [0, 1] *onto the unit square* [0, 1] × [0, 1].

Proof. Intuitively speaking, $f(t)$ is the position at time t of a continuously moving point, beginning at the bottom left corner $\langle 0, 0 \rangle$ of the square at time 0, and ending at the top right corner $\langle 1, 1 \rangle$ at time 1. Thus, a *first approximation* to the curve is a line segment from $\langle 0, 0 \rangle$ to $\langle 1, 1 \rangle$.

We *refine the approximation* by dividing the square into nine equal subsquares, constraining the moving point to spend 1/9 of the unit time interval in each, and traveling from one corner to its opposite in the order shown in Fig. 4.5. Inside the corners in question we show the time at which the moving point arrives. (Notice that certain corners are visited more than once—this is unavoidable.)

We repeat this process in each subsquare, dividing it into nine equal squares and visiting them in zigzag order in nine equal time intervals, and so on. It follows

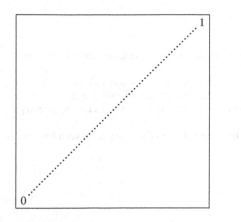 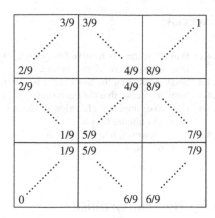

Fig. 4.5 Refining an approximation to the Peano curve

that, for any point P in the square, the moving point visits P at some time t. If P is the limit point of a nested sequence of subsquares, then t is the limit of the corresponding nested sequence of time intervals.

The function

$$f(t) = \text{position of the moving point at time } t$$

is therefore a surjection of $[0, 1]$ onto the unit square. It is also clear that f is continuous for each $t \in [0, 1]$. Given any $\varepsilon > 0$, we can ensure that $f(t')$ is within ε of $f(t) = P$ by finding a subsquare that contains P and is small enough that all its points are within distance ε of P. Then if t' lies within the corresponding subinterval of $[0, 1]$ we have

$$|f(t) - f(t')| < \varepsilon.$$

If t is not on the boundary of the subinterval, call it I, we take

$$\delta = \text{distance from } t \text{ to the nearest end of } I$$

to ensure that

$$|t - t'| < \delta \Rightarrow |f(t) - f(t')| < \varepsilon.$$

If t lies on the boundary of two subintervals I_1, I_2 we take δ to be the minimum of their lengths, and it is again true that

$$|t - t'| < \delta \Rightarrow |f(t) - f(t')| < \varepsilon,$$

because the moving point travels through subsquares of equal size in equal subintervals of time, and we have already arranged that $f(t)$ moves distance less than ε from P in the time intervals in question. $\qquad\square$

Exercises

4.4.1 Prove that the Koch curve has infinite length, by showing that the lengths of the polygons tending to it grow without bound.

4.4.2 Show in fact that the Koch curve has infinite length between any two of its points.

4.4.3 Similarly prove that the Peano curve has infinite length between any two of its points.

4.4.4 Give an example of a function f whose graph $y = f(x)$ for $0 < x < 1$ has a tangent at every point and infinite length.

4.4.5 Show, however, that the curve $y = f(x)$ in Exercise 4.4.4 has finite length between any two of its points.

4.5 Homeomorphisms

In Sect. 3.4.1 we constructed a bijection between the line $\mathbb{R}$ and the plane $\mathbb{R} \times \mathbb{R}$, and it is not hard to modify it so as to obtain a bijection between the line segment $[0, 1]$ and the square $[0, 1] \times [0, 1]$. These bijections are not continuous, but in Sect. 4.4 we have seen a continuous map of $[0, 1]$ onto $[0, 1] \times [0, 1]$. This raises the question whether there is a *continuous bijection* between $[0, 1]$ and $[0, 1] \times [0, 1]$, or between the line and the plane. Such a bijection would be truly disturbing, because it would say that there is essentially no difference between one dimension and two, so the concept of dimension would have no meaning.

The concept of dimension was saved when Brouwer (1911) proved the following theorem on *invariance of dimension*: when $m \neq n$ there is no bijection between $\mathbb{R}^m$ and $\mathbb{R}^n$ that is continuous in both directions. A bijection that is continuous in both directions is called a *homeomorphism*, and the study of properties that are invariant under homeomorphisms is the subject of *topology*. Obviously, topology overlaps with the study of real numbers and continuous functions. But it is a big subject, and we cannot go far into it here. We will be content to prove the simplest case of invariance of dimension:

Distinctness of dimensions one and two. *There is no continuous bijection between* $\mathbb{R}$ *and* $\mathbb{R}^2$.

Proof. We use a property of $\mathbb{R}$ that it does not share with $\mathbb{R} \times \mathbb{R}$; namely, $\mathbb{R}$ *can be separated by a point.* To be precise, if we remove the point 0, the points 1 and -1 in the resulting set $\mathbb{R} - \{0\}$ cannot be joined by a continuous path. Indeed, a continuous path from -1 to 1 is a continuous function $f : [0, 1] \to \mathbb{R} - \{0\}$ with $f(0) = -1$ and $f(1) = 1$. Such a function fails to satisfy the intermediate value theorem, because it does not take the value 0.

Now if there is a bijection $g : \mathbb{R} \to \mathbb{R}^2$, continuous in both directions, consider the points $g(-1)$, $g(0)$, and $g(1)$. These are three distinct points in the plane, so there is a continuous path from $g(-1)$ to $g(1)$ that does not meet $g(0)$. If we transport this path back to $\mathbb{R}$ by the continuous bijection g^{-1} we get a continuous path from -1 to 1 not meeting 0.

As we have just seen, such a path is impossible in $\mathbb{R}$, so the continuous bijection g does not exist.   □

Exercises

In a striking contrast to the topological distinctness of $\mathbb{R}$ and $\mathbb{R} \times \mathbb{R}$, there is no such distinction between C and $C \times C$, where C is the Cantor set. Given any $x \in C$, we separate its ternary expansion into the sequence x_1 of odd-position digits and the sequence x_2 of even-position digits. For example, if

$$x = (0.022022022022\ldots)_3$$

then

$$x_1 = (0.0\ 2\ 2\ 0\ 2\ 2\ \ldots)_3$$

and

$$x_2 = (0.\ 2\ 0\ 2\ 2\ 0\ 2\ldots)_3$$

4.5.1   Explain why the map $x \mapsto \langle x_1, x_2 \rangle$ is a bijection of C onto $C \times C$.
4.5.2   Also explain why the map $x \mapsto \langle x_1, x_2 \rangle$ is continuous and has a continuous inverse.

It is not always the case that a continuous bijection has a continuous inverse.

4.5.3   Give an example of a bijection between $[0,1)$ and the circle that is continuous in one direction but not in the other.

4.6   Uniform Convergence

The continuous Peano function f of Sect. 4.4 can be viewed as the limit of a sequence of very simple continuous functions $f_1, f_2, f_3, \ldots$. Each f_i maps the unit interval into a polygonal path through the unit square, zigzagging through the diagonals of subsquares of width $1/3^i$.

The sequence $f_1, f_2, f_3, \ldots$ not only converges to f, it converges *uniformly* in the following sense:

Definition. Functions f_n *converge uniformly* to f if, for any $\varepsilon > 0$, we can find an N such that

$$n > N \Rightarrow |f_n(t) - f(t)| < \varepsilon \quad \text{for all } t.$$

In other words, $f_n(t)$ and $f(t)$ differ by less than ε *at all points t*.

This property is clear for the functions f_n that converge to the Peano curve. If $n > N$, then the polygonal path defined by f_n is a refinement of the path defined by f_N, and it falls within the same zigzag sequence of squares traversed by f_N.

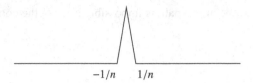

$-1/n$ $1/n$

Fig. 4.6 The spike function

Therefore, f_n and f differ from each other by at most the diameter of these squares, which can be made as small as we please by choosing N sufficiently large.

This idea gives the following criterion for the limit of a sequence of continuous functions to be continuous.

Uniform Convergence Criterion. *If $f_1, f_2, f_3, \ldots$ is a uniformly convergent sequence of continuous functions on an interval $[a, b]$, then*

$$f(x) = \lim_{n \to \infty} f_n(x)$$

is also continuous.

Proof. For each $c \in [a, b]$ we wish to show that $\lim_{x \to c} f(x) = f(c)$. That is, for each $\varepsilon > 0$ we seek a $\delta > 0$ such that

$$|x - c| < \delta \Rightarrow |f(x) - f(c)| < \varepsilon.$$

We can do this by finding an N and a δ for which the following three conditions hold simultaneously:

1. $|f(x) - f_n(x)| < \varepsilon/3$ for $n > N$,
2. $|f_n(x) - f_n(c)| < \varepsilon/3$ for $|x - c| < \delta$,
3. $|f_n(c) - f(c)| < \varepsilon/3$ for $n > N$.

Conditions 1 and 3 can be met simultaneously by uniform convergence of the sequence $f_1, f_2, f_3, \ldots$. Condition 2 can be met, by the continuity of f_n, for some δ depending on N and c.

So, if we first choose N to meet conditions 1 and 3, then choose δ to meet condition 2, we have a δ such that $|x - c| < \delta \Rightarrow |f(x) - f(c)| < \varepsilon$, as required.   □

Without the condition of uniform convergence a convergent sequence of continuous functions may have a discontinuous limit. For example, let $f_n(x)$ be the function with the spike-shaped graph shown in Fig. 4.6.

That is,

$$
f_n(x) = \begin{cases}
0 & \text{if } x \leq -1/n \\
nx + 1 & \text{if } -1/n \leq x \leq 0 \\
-nx + 1 & \text{if } 0 \leq x \leq 1/n \\
0 & \text{if } x \geq 1/n.
\end{cases}
$$

It is clear, since the spike becomes arbitrarily thin as n increases, that

$$
f(x) = \lim_{n \to \infty} f_n(x) = \begin{cases}
0 & \text{if } x < 0 \\
1 & \text{if } x = 0 \\
0 & \text{if } x > 0,
\end{cases}
$$

which is discontinuous at $x = 0$. The discontinuity arises from the nonuniform convergence of the f_n to f. In particular, there is no N for which

$$
n > N \Rightarrow |f_n(x) - f(x)| < 1/2 \text{ for all } x,
$$

because we always have $|f_n(1/2n) - f(1/2n)| = 1/2$.

Exercises

4.6.1 If $g_1, g_2, g_3, \ldots$ are the continuous functions defining the first, second, third, ... polygonal approximations to the Koch curve given in Sect. 4.4, show that the sequence $g_1, g_2, g_3, \ldots$ converges uniformly.

4.6.2 Give an example of a function with infinitely many discontinuities that is the (nonuniform) limit of a sequence of continuous functions.

4.7 Uniform Continuity

Any continuous function on a closed interval $[a, b]$ is actually continuous in the "uniform" sense exhibited by the Peano curve: the variation of $f(x)$ can be kept within ε by keeping the variation of x within some δ *which does not depend on x.* The formal definition of uniform continuity can be stated as follows.

Definition. A function f is *uniformly continuous* on a set S if, for any $\varepsilon > 0$ and any $x, y \in S$, there is a $\delta > 0$ such that

$$
|x - y| < \delta \Rightarrow |f(x) - f(y)| < \varepsilon.
$$

We notice that a uniformly continuous function is continuous at each point of $c \in S$. Because, if we fix $y = c$, we have

$$|x - c| < \delta \Rightarrow |f(x) - f(c)| < \varepsilon,$$

so $\lim_{x \to c} f(x) = f(c)$.

However, continuity does not always imply uniform continuity. The function $f(x) = 1/x$ is continuous on the set $S = (0, 1)$ but *not* uniformly continuous, because we can have $|f(x) - f(y)| \geq 1$ while $|x - y|$ is as small as we please, by choosing x and y sufficiently small.

The concepts of continuity and uniform continuity agree on *closed intervals* S, thanks to the following theorem.

Uniform continuity on closed intervals. *A continuous function on a closed interval is uniformly continuous.*

Proof. Suppose f is continuous on $[a, b]$. Since f is continuous at each $c \in [a, b]$, for each $\varepsilon > 0$ there is a $\delta(c)$ such that

$$|x - c| < \delta(c) \Rightarrow |f(x) - f(c)| < \varepsilon/2.$$

It follows that

$$x, y \in (c - \delta(c), c + \delta(c)) \Rightarrow |f(x) - f(y)| < \varepsilon,$$

because $|f(x) - f(c)| < \varepsilon/2$ and $|f(y) - f(c)| < \varepsilon/2$. Thus, each $c \in [a, b]$ lies in an open interval U with the property that

$$x, y \in U \Rightarrow |f(x) - f(y)| < \varepsilon. \tag{*}$$

The set of all open intervals U with property (*) therefore covers $[a, b]$. It follows, by the Heine–Borel theorem of Sect. 3.6, that $[a, b]$ is covered by *finitely many* open intervals $U_1, U_2, \ldots, U_n$, each with property (*).

Let $c_1 < c_2 < \cdots < c_m$ be the endpoints of $U_1, U_2, \ldots, U_n$ lying between a and b. Also let $a = c_0$ and $b = c_{m+1}$. Since each $c_i \in$ some U_k, the open intervals on either side of c_i, (c_{i-1}, c_i) and (c_i, c_{i+1}), are contained in U_k. So if we let

$$\delta = \text{minimum length among the intervals } (c_i, c_{i+1}),$$

we have

$$|x - y| < \delta \Rightarrow x, y \in \text{ same } U_k \Rightarrow |f(x) - f(y)| < \varepsilon,$$

as required to show that f is uniformly continuous on $[a, b]$. $\qquad\qquad\square$

This proof also has the following more general consequence, for the *compact* sets introduced in Sect. 3.6. The consequence will become useful when we obtain a clearer view of compact sets in Sect. 5.4.

Corollary 2. *A continuous function on a compact set is uniformly continuous.*

Proof. Suppose f is continuous on a compact set K. For each $x_0 \in K$ and each $\varepsilon > 0$ continuity implies there is a δ such that

$$|x - x_0| < \delta \Rightarrow |f(x) - f(x_0)| < \varepsilon/2.$$

Taking all such x_0 and δ, we get a covering of K by the open intervals $(x_0 - \delta, x_0 + \delta)$. By compactness, finitely many of these intervals also cover K. We can then argue as in the proof above. $\square$

Exercises

Since a curve with endpoints is a continuous function on a compact set, namely $[0,1]$, we can deduce some general properties of such curves from uniform continuity.

4.7.1 For any curve in the plane $f : [0.1] \rightarrow \mathbb{R}^2$ and any $\varepsilon > 0$ show that there are values $t_1, t_2, \ldots, t_k \in [0, 1]$ such that the section of the curve between $f(t_i)$ and $f(t_{i+1})$ lies within a circle of radius ε.

4.7.2 Deduce from Exercise 4.7.1 that any curve with endpoints is the uniform limit of a sequence of polygons.

Suppose we want to define curves *without* endpoints, in order to cover curves such as the parabola.

4.7.3 Propose a suitable definition, involving the concept of a continuous function.

4.7.4 Illustrate your definition in the case of the parabola.

4.8 The Riemann Integral

The concept of uniform continuity fits like a glove onto the concept of Riemann integral used in basic calculus. Recall that the definition of $\int_a^b f(x)\, dx$ involves the following setup, as shown in Fig. 4.7:

1. A closed interval $[a, b]$ on which $f(x)$ is defined.
2. A division of $[a, b]$ by finitely many points c_i with $a < c_1 < c_2 < \cdots < c_m < b$.
3. Lower and upper approximations to f by step functions, constant on each interval (c_i, c_{i+1}). The lower approximation equals the minimum m_i of f on each interval, and the upper approximation equals the maximum, M_i.
4. Lower and upper approximations to the integral, $\sum_i m_i(c_{i+1} - c_i)$ and $\sum_i M_i(c_{i+1} - c_i)$, called *Riemann sums*.

5. The integral exists if the difference between the two Riemann sums (made of rectangles like the one shown shaded in Fig. 4.7) can be made arbitrarily small.

A uniformly continuous f is tailor-made for this setup, because we can obtain finitely many intervals on which the difference between the minimum and maximum of f is less then ε. Consequently the difference in area between the upper and lower approximations, is less than $(b-a)\varepsilon$, which can be made arbitrarily small. Therefore, since a continuous function on a closed interval is uniformly continuous, we have the theorem:

Integrability of continuous functions. *If f is continuous on $[a, b]$ then $\int_a^b f(x)\, dx$ exists.*   □

However, there is not a perfect match between continuous functions and Riemann integrable functions, because certain *dis*continuous functions are also Riemann integrable. An easy example is the function

$$f(x) = \begin{cases} 1 \text{ if } x = 0 \\ 0 \text{ if } x \neq 0, \end{cases}$$

the integral of which equals 0 on any interval. This is because the lower approximation is the constant zero function, and the upper approximation can be taken to be zero everywhere except on an arbitrarily small interval $(-\varepsilon, \varepsilon)$.

In fact, a function with a dense set of discontinuities can be Riemann integrable (see exercises), though not all such functions are. The Dirichlet function mentioned in Sect. 4.2 is not Riemann integrable. Because of this, a more general concept of integrability is desirable. The best-known general integral, the *Lebesgue integral*, is based on a general concept of *measure*, which will be discussed in Chap. 9. One of

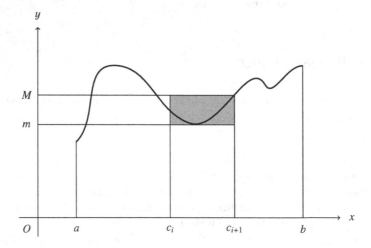

Fig. 4.7 Setup for the Riemann integral

the beauties of Lebesgue measure, besides its generality, is that it gives a precise criterion for a bounded function to be Riemann integrable. Namely, *a bounded f is Riemann integrable if and only if f is continuous everywhere except on a set of measure zero.* We prove this result in Sect. 9.5.

4.8.1   The Fundamental Theorem of Calculus

The fundamental theorem of calculus, roughly speaking, states that the derivative of the integral of a function f equals f. Various versions of the theorem exist, depending on the nature of the integral and the functions f. The simplest version, which we will now prove, concerns the Riemann integral of a continuous function f.

Fundamental theorem of calculus. *If f is continuous and $F(x)$ is the Riemann integral $\int_a^x f(t)\,dt$, then $F'(x) = f(x)$.*

Proof. By the definition of derivative,

$$F'(x) = \lim_{n \to 0} \frac{F(x+h) - F(x)}{h}$$

$$= \lim_{h \to 0} \frac{\int_a^{x+h} f(t)\,dt - \int_a^x f(t)\,dt}{h}$$

$$= \lim_{h \to 0} \frac{1}{h} \int_x^{x+h} f(t)\,dt.$$

Now it follows from the definition of Riemann integral that

$$hm \le \int_x^{x+h} f(t)\,dt \le hM,$$

where m and M are the minimum and maximum of $f(t)$ for $h \in [x, x+h]$. Consequently,

$$m \le \frac{1}{h} \int_x^{x+h} f(t)\,dt \le M,$$

and as $h \to 0$ both $m, M \to f(x)$ by the continuity of f. Thus,

$$F'(x) = \lim_{h \to 0} \frac{1}{h} \int_x^{x+h} f(t)\,dt = f(x). \qquad \square$$

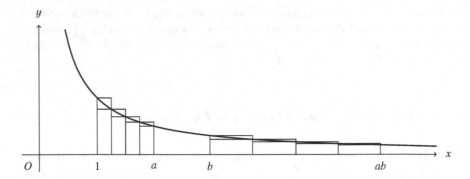

Fig. 4.8 Riemann sums from 1 to a and from b to ab

Exercises

The defining property of the logarithm, $\log ab = \log a + \log b$, can be proved directly from the definition of $\log c$ as a Riemannn integral. We define

$$\log c = \int_1^c \frac{dx}{x}$$

and consider the integral of $1/x$ from 1 to a and from b to ab. This amounts to comparing Riemann sums for the two areas shown under the curve in Fig. 4.8. We use Riemann sums obtained by dividing both $[0, a]$ and $[b, ab]$ into n equal parts (Fig. 4.8 shows the case $n = 4$).

4.8.1 Show that the rectangles in the Riemann sums from 1 to a have exactly the same areas as their counterparts in the Riemann sums from b to ab.

4.8.2 Deduce from Exercise 4.8.1 that $\int_1^a \frac{dx}{x} = \int_b^{ab} \frac{dx}{x}$.

4.8.3 Deduce from Exercise 4.8.2 that $\log ab = \log a + \log b$.

Now we show that the Thomae function $t(x)$ defined in the exercises to Sect. 4.2, despite its many discontinuities, is Riemann integrable. The proof depends on the fact that unequal intervals are allowed.

4.8.4 By subdividing $[0,1]$ into unequal subintervals in such a way that "large" values of $t(x)$ are enclosed in "narrow" subintervals, show that the Riemann sums for $t(x)$ can be made arbitrarily small.

4.8.5 Deduce that $\int_0^1 t(x)\, dx = 0$.

4.8.6 Show, on the other hand, that $t(x)$ does not satisfy the fundamental theorem of calculus. Namely, if $F(x) = \int_0^x t(x)\, dx$ then $F'(x) = t(x)$ only for irrational x.

4.9   Historical Remarks

From its beginnings in the seventeenth century, calculus was supposed to deal with continuous phenomena. For Newton, in particular, the basic phenomenon was continuous *motion*, and the basic problems were the following two:

1. Given the length of space continuously (that is, at every time), to find the speed of motion at any time proposed.
2. Given the speed of motion continuously, to find the length of space described at any time proposed.

Newton (1671), p. 71.

In our language, these are the problems of differentiating and integrating continuous functions. In Problem 1 we are given distance $d(t)$ and have to find the speed $d'(t)$. In Problem 2 we are given the speed $v(t)$ and have to find the distance traveled by time T, $\int_0^T v(t)\, dt$. The mental picture of continuous motion makes it plausible that $d'(t)$ exists for any continuous distance function $d(t)$, and that $\int_0^T v(t)\, dt$ exists for any continuous speed function $v(t)$. But, as we now know, only the second of these statements is true.

In fact, as calculus evolved, the notion of continuous function expanded, from being identical with the notion of differentiable function until it included functions that are *nowhere* differentiable. The shift in meaning was partly due to expansion of the function concept, and partly due to the tardy development of the limit concept, without which a precise definition of continuity was not possible.

As we saw in Sects. 1.5 and 1.9, functions were originally dependent on formulas, and the first formulas considered were indeed differentiable. The concept of "formula" expanded with the discovery of Fourier series, which could express functions that were clearly not differentiable at all points. Take the triangular wave function of Sect. 1.5, for example, which clearly has infinitely many "corners" at which no tangent exists.

Bolzano (1817) first gave a definition via continuity at a point, essentially as we do today, and Cauchy (1821) rediscovered the concept in the first comprehensive and rigorous course on analysis. Cauchy's course included precise concepts of limit, convergence of series, and continuity, and also a concept of integral that enabled him to prove that every continuous function (on a closed interval) is integrable. At this time it was still believed that continuous functions are differentiable, except perhaps at a "few" exceptional points. The incentive to study more "pathological" continuous functions came from the theory of Fourier series.

Fourier (1822) discovered that, under certain conditions, a function f could be expressed in the form

$$f(x) = \frac{1}{2} + \sum_{n=1}^{\infty} (a_n \cos nx + b_n \sin nx),$$

where the coefficients are the integrals

$$a_n = \frac{1}{\pi} \int_{-\pi}^{\pi} f(x) \cos nx \, dx \quad \text{and} \quad b_n = \frac{1}{\pi} \int_{-\pi}^{\pi} f(x) \sin nx \, dx.$$

The conditions for these formulas to be valid were not clear, and Dirichlet (1829) was the first to prove a general theorem. He showed that the formulas for a_n and

Fig. 4.9 Augustin-Louis Cauchy and Peter Gustav Lejeune Dirichlet

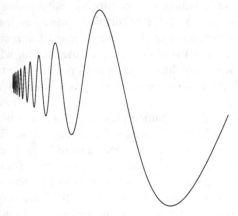

Fig. 4.10 A continuous function with infinitely many oscillations

b_n are valid provided that f on $(-\pi, \pi)$ is continuous and *piecewise monotonic*. The latter condition means that $(-\pi, \pi)$ can be divided into finitely many subintervals, on each of which f is either nondecreasing or nonincreasing.

Thus, the validity of Fourier series was not yet proved for continuous functions with infinitely many oscillations, such as $f(x) = x \sin \frac{1}{x}$ for $x > 0$ and $f(x) = 0$ for $x = 0$. (Fig. 4.10).

This led to the investigation of wildly oscillating continuous functions, and eventually to continuous counterexamples to Fourier's formulas. More importantly, it led to the discovery of nowhere differentiable functions by Weierstrass (1872). The first examples were based on infinite sums of trigonometric functions and were hard to visualize [though see Hairer and Wanner (1996), p. 265, for an

Fig. 4.11 Giuseppe Peano and Helge von Koch

Fig. 4.12 Example of a simple closed curve

understandable example]. The von Koch (1904) example became famous because of its visual appeal.

The Koch curve and the Peano curve showed how far the notion of continuous curve had evolved since the seventeenth century. By the end of the nineteenth century one had a simple and general definition of a continuous curve, but the definition covered examples more pathological (and interesting?) than originally intended. Still, the definition passed one important test for continuous curves: the *Jordan curve theorem*. This theorem, formulated by Jordan (1887) states that *a simple* (that is, not self-intersecting) *closed curve separates the plane into two regions* (the "inside" and "outside" of the curve). The Jordan curve theorem is correct, but hard to prove. The proof by Jordan was considered suspect by his successors, but was declared essentially correct by Hales (2007), who produced the first computer-checkable proof of the theorem.

Some beautiful examples of simple closed curves (which hint at why the Jordan curve theorem may be hard to prove) have recently been given in the field of *TSP art*. See, for example, the web site of Robert Bosch

www.oberlin.edu/math/faculty/bosch/tspart-page.html

Figure 4.12 is one of his images.

Chapter 5
Open Sets and Continuity

PREVIEW

In this chapter we shift our attention from functions back to sets. The shift is prompted by the fact that continuous functions have a natural description in terms of sets: the so-called *open* sets. Just as continuous functions may be viewed as the simplest functions, open sets may be viewed as the simplest sets. And just as complicated functions arise from continuous functions by the limit process, complicated sets arise from open sets by certain operations, namely, complementation and countable union.

There are in fact parallel classifications of functions and sets into levels of complexity, called the *Baire hierarchy* of functions and the *Borel hierarchy* of sets. Both are useful, and they interact usefully with each other, but sets are a little easier to work with. We study the classification of Borel sets in depth in Chap. 8.

Here we begin studying the classification by looking at open sets and their complements, the *closed* sets. After describing open sets and their relationship with continuous functions, we introduce closed sets and focus on a particular type, the *perfect* sets. An example of a perfect set is the Cantor set C, and we show that every perfect set resembles C in a certain sense.

Finally, we lay the foundation for an orderly construction of the Borel sets by constructing a *universal open set*—an open set in $\mathbb{R}^2$ whose horizontal sections are precisely the open sets in $\mathbb{R}$. More precisely, we do this with a convenient replacement for $\mathbb{R}$, the set $\mathcal{N}$ of all functions $f : \mathbb{N} \to \mathbb{N}$. $\mathcal{N}$ can be viewed as the set of irrational numbers in $(0,1)$, and it avoids problems of ambiguity caused by the rational numbers.

5.1 Open Sets

An open interval on the line $\mathbb{R}$ is a special case of the concept of *open set*. The concept of open set is actually very general but for our purposes a set U is open if,

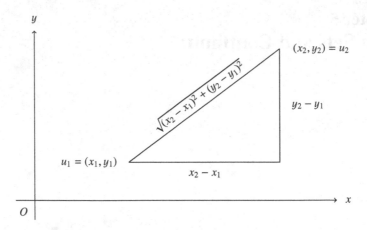

Fig. 5.1 Distance in $\mathbb{R}^2$

Fig. 5.2 An ε-neighborhood in $\mathbb{R}^2$

for each point $u \in U$, the points within some positive distance ε of u also belong to U. Thus, our concept of open set applies to spaces with a concept of *distance*; typically the n-dimensional Euclidean spaces $\mathbb{R}^n$.

For example, in the plane $\mathbb{R}^2$, the distance from point $u_1 = (x_1, y_1)$ to point $u_2 = (x_2, y_2)$ is given by

$$|u_2 - u_1| = \sqrt{(x_2 - x_1)^2 + (y_2 - y_1)^2},$$

thanks to the Pythagorean theorem; see Fig. 5.1. The set of points at distance less than ε from $u \in \mathbb{R}^2$ is an *open disk* of radius ε and center u, shown in Fig. 5.2. (The boundary circle is drawn dashed to indicate that it does not belong to the neighborhood.)

This concept of distance easily generalizes to $\mathbb{R}^n$, where the set of points at distance less than ε from a point u, $N_\varepsilon(u) = \{v \in \mathbb{R}^n : |v - u| < \varepsilon\}$, is called an *open n-ball* of radius ε or the *ε-neighborhood* of u.

Given the concept of ε-neighborhood, we can define the concept of an open set as follows.

Definition. A set $U \subseteq \mathbb{R}^n$ is called *open* if, for each point $u \in U$, there is an $\varepsilon > 0$ for which $N_\varepsilon(u)$ is contained in U.

The basic properties of open sets follow easily from this definition.

1. The empty set (trivially) and the whole space $\mathbb{R}^n$ are open sets.
2. The union of any collection of open sets is open.
3. The intersection $U_1 \cap U_2$ of two open sets U_1 and U_2 is open.

   Because, if $u \in U_1 \cap U_2$ there is an ε_1 for which $N_{\varepsilon_1}(u) \subseteq U_1$ and an ε_2 for which $N_{\varepsilon_2}(u) \subseteq U_2$. So if we take $\varepsilon = \min(\varepsilon_1, \varepsilon_2)$ we have ε for which $N_\varepsilon(u) \subseteq U_1 \cap U_2$, as required.

4. Each open U set is the union of open n-balls.

   In particular, U is the union of the set of n-balls $N_\varepsilon(u)$ for the $u \in U$ and the ε (depending on u) for which $N_\varepsilon(u) \subseteq U$.

5. In fact, each open set U is the union of *countably many* n-balls.

   It suffices to take n-balls of rational radius centered on points u with rational coordinates. Any point $v \in U$ has a neighborhood $N_\varepsilon(v) \subseteq U$, and inside $N_\varepsilon(v)$ we can take u with rational coordinates $\langle r_1, r_2, \ldots, r_n \rangle$ so close to v that $v \in N_r(u)$ for a rational r small enough to ensure that $N_r(u) \subseteq N_\varepsilon(v) \subseteq U$.

   Then, as we know from Sect. 3.1, the set of all $(n+1)$-tuples $\langle r, r_1, r_2, \ldots, r_n \rangle$ of members of a countable set is countable.

6. It follows from property 5 that the set of open sets $U \subseteq \mathbb{R}^n$ is equinumerous with $\mathbb{R}$.

   This follows from the result of Sect. 3.3 that the set of subsets of a countable set (which we took to be subsets of $\mathbb{N}$ in that section) is equinumerous with $\mathbb{R}$.

Even in $\mathbb{R}$, open sets can be quite complex and interesting. An example is the complement[1] $[0, 1] - C$ of the Cantor set in the unit interval, introduced in Sect. 3.7. In this case we can explicitly list countably many open intervals whose union is $[0, 1] - C$, namely:

$$\left(\tfrac{1}{3}, \tfrac{2}{3}\right)$$

$$\left(\tfrac{1}{9}, \tfrac{2}{9}\right) \qquad\qquad \left(\tfrac{7}{9}, \tfrac{8}{9}\right)$$

$$\left(\tfrac{1}{27}, \tfrac{2}{27}\right) \quad \left(\tfrac{7}{27}, \tfrac{8}{27}\right) \qquad \left(\tfrac{19}{27}, \tfrac{20}{27}\right) \quad \left(\tfrac{25}{27}, \tfrac{26}{27}\right)$$

and so on.

Exercises

The open disks or balls are often called *basic* open sets because all open sets are obtainable from them as unions. In $\mathbb{R}^n$ for $n > 1$ another useful family of basic open sets are the *cartesian products* of n intervals. For example, in $\mathbb{R}^2$ these cartesian products are *open rectangles*

$$(a, b) \times (c, d) = \{\langle x, y \rangle : a < x < b \text{ and } c < y < d\}.$$

---

[1]Remember that in this book we use the ordinary minus sign to denote set difference.

These basic sets are sometimes more convenient (e.g., in measure theory—see Chap. 9), so it is worth checking that they give the same open sets as the open disks.

5.1.1 Explain why any open rectangle is a union of open disks.
5.1.2 Explain why any open disk is a union of open rectangles.
5.1.3 Explain why any open set $U \subseteq \mathbb{R}^2$ is a countable union of open rectangles.

The complement of an open set is generally not open.

5.1.4 Explain why C is not open.
5.1.5 Show that $\mathbb{R}$ and the empty set are the only open subsets of $\mathbb{R}$ with open complements.

5.2 Continuity via Open Sets

When open sets are defined by ε-neighborhoods, as in the previous section, they are very naturally aligned with the concept of continuous function. In fact, they enable us to define the concept of a continuous function globally, without recourse to the preliminary definition of "continuity at a point."

Continuity in terms of open sets. *A function $f : \mathbb{R} \to \mathbb{R}$ is continuous if and only if $f^{-1}(U)$ is open for each open set U, where*

$$f^{-1}(U) = \{x \in \mathbb{R} : f(x) \in U\}.$$

Proof. By the ε-δ definition of continuity, for each $a \in \mathbb{R}$ and each $\varepsilon > 0$ there is a $\delta > 0$ such that

$$|x - a| < \delta \Rightarrow |f(x) - f(a)| < \varepsilon.$$

In terms of neighborhoods, this says

$$x \in N_\delta(a) \Rightarrow f(x) \in N_\varepsilon(f(a)).$$

And in terms of f^{-1}: for each $a \in \mathbb{R}$ and $\varepsilon > 0$ there is a δ with

$$N_\delta(a) \subseteq f^{-1}(N_\varepsilon(f(a))). \tag{*}$$

Now if U is any open set, and if $a \in f^{-1}(U)$, then $f(a) \in U$. Since U is open, it contains some $N_\varepsilon(f(a))$, and so $f^{-1}(N_\varepsilon(f(a)))$ contains $N_\delta(a)$. In other words, if $a \in f^{-1}(U)$, then $f^{-1}(U)$ also contains some $N_\delta(a)$. That is, $f^{-1}(U)$ is open.

Conversely, if any $f^{-1}(\text{open})$ is open, then $f^{-1}(N_\varepsilon(f(a)))$ is an open set that includes a, and hence some $N_\delta(a)$. This gives us a δ for each ε. $\square$

The ε-δ definition of continuity applies to functions $f : \mathbb{R}^m \to \mathbb{R}^n$, if we interpret $|x - a|$ as the distance between x and a in $\mathbb{R}^m$, and $|f(x) - f(a)|$ as the distance between $f(x)$ and $f(a)$ in $\mathbb{R}^n$. Then the argument given above shows that $f : \mathbb{R}^m \to \mathbb{R}^n$ is continuous if and only if $f^{-1}(U)$ is open (as a subset of $\mathbb{R}^m$) for each open set U in $\mathbb{R}^n$.

5.2.1 The General Concept of Open Set

In this book we are concerned only with spaces in which a concept of distance may be defined, and hence with open sets and continuous functions defined by means of ε-neighborhoods. However, now that we have seen continuity defined in terms of open sets, it is worth mentioning that open sets can be defined *without* use of the concept of distance. The trick is to use the first three properties of open sets, given in Sect. 5.1, as a *definition*.

Suppose that we have a set X and a collection $\mathcal{T}$ of sets $U \subseteq X$ with the following properties.

1. The empty set and X itself are members of $\mathcal{T}$.
2. The union of any members of $\mathcal{T}$ is a member of $\mathcal{T}$.
3. The intersection of any two members of $\mathcal{T}$ is a member of $\mathcal{T}$.

Then the sets $U \in \mathcal{T}$ are called *open* subsets of X, and $\mathcal{T}$ is called a *topology* on X. Using this concept of an open set, we can talk about continuous functions, homeomorphisms, and so on, without depending on a concept of distance. Any other concepts that can be defined in terms of open sets (such as closed sets and compact sets—see below) are also meaningful in this general theory of open sets, which is called *general topology*.

Exercises

The open set definition of continuity avoids having to define continuity at a point, but it is also easy to define continuity at a point in terms of open sets.

5.2.1 Express the continuity of f at point a in terms of open sets containing the point $f(a)$.

Any set $S \subseteq \mathbb{R}$ has an obvious topology, called the *relative topology*, whose open sets are precisely the sets $S \cap U$, where U is an open subset of $\mathbb{R}$.

5.2.2 Check that the sets $S \cap U$ satisfy the three conditions for a topology.

5.2.3 Show that the relative topology on $[0,1]$ has basic open sets of the form $[0, b)$, (a, b), and $(a, 1]$, for $a, b \in (0, 1)$.

In the case where $S = C$, the relative topology is particularly interesting, because it has a countable collection of basic open sets that arise naturally from inside C. We recall from Sect. 3.7 that the elements of C are those numbers in C with ternary expansions that can be written using only the digits 0 and 2.

5.2.4 Consider ternary expansions of the form

$$x = 0.02a_1a_2a_3 \ldots \quad \text{where} \quad a_1, a_2, a_3, \ldots = 0 \text{ or } 2.$$

Show that the numbers of this form make up the intersection of C with the interval $[2/9, 1/3]$, which is also the intersection of C with the open interval $(2/9 - \varepsilon, 1/3 + \varepsilon)$ for ε sufficiently small.

5.2.5 By generalizing the idea of Exercise 5.2.4, show that, for any sequence $b_1 b_2 \cdots b_k$ of digits $b_i = 0$ or 2, the set

$$F(b_1, \ldots, b_k) = \{x : x = b_1 \ldots b_k a_1 a_2 a_3 \ldots \text{ for some } a_1, a_2, a_3, \ldots = 0 \text{ or } 2\}$$

is an open set in the relative topology of C.

5.2.6 Show also that the sets $F(b_1, \ldots, b_k)$ are basic open sets in the relative topology for C.

It is a similar story for the set $\mathcal{N}$ of irrational numbers in $[0,1]$. We know from Sects. 2.7 and 2.8 that $\mathcal{N}$ can be identified (via continued fractions) with the infinite sequences $\langle a_1, a_2, a_3, \ldots \rangle \in \mathbb{N}^{\mathbb{N}}$.

5.2.7 Show that the set

$$G(b_1, \ldots, b_k) = \{x : x = \langle b_1, \ldots, b_k, a_1, a_2, \ldots \rangle \text{ for some } a_1, a_2, \ldots \in \mathbb{N}\}$$

is an open set in the relative topology on $\mathcal{N}$.

5.2.8 Show also that the sets $G(b_1, \ldots, b_k)$ are basic open sets in the relative topology for $\mathcal{N}$.

5.3   Closed Sets

The complement $\mathbb{R}^n - U$ of an open set U in $\mathbb{R}^n$ is called *closed*. For example, a closed interval $[a, b]$ in $\mathbb{R}$ is a closed set because its complement $(-\infty, a) \cup (b, \infty)$ is open. The basic properties of closed sets follow from those of open sets, enumerated in Sect. 5.1, by taking complements. In particular:

1. The empty set and the whole space $\mathbb{R}^n$ are closed.
2. An arbitrary intersection of closed sets is closed.

   If the closed sets F_i in question are the complements of open sets U_i, then

$$\text{intersection of the } F_i = \text{complement(union of the } U_i)$$

$$= \text{complement of an open set}$$

$$= \text{closed set.}$$

3. The union of two closed sets is closed.

   If the closed sets F_1 and F_2 are the complements of U_1 and U_2, respectively, then

$$F_1 \cup F_2 = \text{complement}(U_1 \cap U_2)$$

$$= \text{complement of an open set}$$

$$= \text{closed set.}$$

4. The set of closed sets is equinumerous with $\mathbb{R}$.

   Because the complement operation gives a bijection between the set of closed sets and the sets of open sets, which we know is equinumerous with $\mathbb{R}$.

F is traditionally used to denote closed sets because it is the initial letter of the French word *fermé*, meaning closed. On the same grounds, one might expect O to be used to denote open sets, because the French word for open is *ouvert*. However, the letter O is likely to be confused with 0, which is probably why one uses the second letter, U, instead.

Note that "closed" does *not* mean "not open," because there are many sets that are neither open nor closed. A closed set is "closed" in the sense that it includes all its limit points.

Closure of Closed Sets. *If F is a closed set and x is a limit point of F, then $x \in F$. Conversely, any set that includes all its limit points is closed.*

Proof. Recall from Sect. 3.6 that x is a limit point of F if every ε-neighborhood of x includes points of F other than x.

It follows that x is not in the complement of F, because the complement of F is open and hence contains an ε-neighborhood of each of its points.

Conversely, suppose F is a set that includes all of its limit points. Then each point $y \notin F$ is *not* a limit point of F, so y has an open neighborhood disjoint from F. In other words, the complement of F contains an open neighborhood of each of its points, and hence is open, so F is closed. $\square$

Exercises

5.3.1 Show that $\mathbb{Q}$ is neither open nor closed in $\mathbb{R}$.

5.3.2 Show that, for $a < b$, the half-open interval $[a, b) = \{x : a \le x < b\}$ is neither open nor closed.

5.3.3 Show that the examples in Exercises 5.3.1 and 5.3.2 are countable unions of closed sets.

On the other hand, in certain topologies many sets are both open *and* closed.

5.3.4 Show that the complement of the basic open set $F(0, 2)$ in C equals the open set $F(0, 0) \cup F(2)$, so $F(0, 2)$ is closed.

5.3.5 More generally, show that the complement of any basic open set in C is a finite union of basic open sets.

5.3.6 Show similarly that the complement of any basic open set in $\mathcal{N}$ is a countable union of basic open sets.

5.4   Compact Sets

In Sect. 3.6 we proved the Heine–Borel theorem about covering $[0,1]$ by open intervals. We remarked that the Heine–Borel property ("arbitrary cover contains a finite subcover") is the defining property of what we now call *compact* sets. We now replay the proof of the Heine–Borel theorem to prove the following:

Characterization of compact sets in $\mathbb{R}$. *A set $K \subseteq \mathbb{R}$ is compact if and only if it is closed and bounded.*

Proof. First suppose that K is closed and bounded.

Since K is bounded, we can "bisect" it by bisecting the interval between an upper and lower bound of K, and then we can proceed as in the proof of the Heine–Borel theorem. That is, we suppose K is covered by an infinite set of intervals U_i, with no finite subcover, and repeatedly choose a "half" with no finite subcover. In this way we obtain a nested sequence of intervals $I_1 \supset I_2 \supset I_3 \supset \cdots$ with the following properties.

1. Each I_{n+1} is half the length of I_n.
2. Each I_n contains points of K.
3. The set $I_n \cap K$ is covered by infinitely many of the intervals U_i, but not by finitely many of them.

This implies that $\bigcap_{n=1}^{\infty} I_n$ is a single point x, which belongs to the closed set K because it is a limit point of K. It follows that $x \in U_j$ for some open interval U_j in the collection covering K. But then $I_k \subset U_j$ for sufficiently large k, contradicting the assumption that $I_k \cap K$ cannot be covered by finitely many of the intervals U_i.

This contradiction establishes that a closed bounded set is compact.

Conversely, suppose that K is compact.

If K is unbounded then we can cover it by the intervals $U_n = (-n, n)$, but not by finitely many of these U_n; hence K must be bounded. If x is a limit point of K but $x \notin K$ then we can cover K by the intervals

$$V_n = \left(-\infty, x - \frac{1}{n}\right) \quad \text{and} \quad W_n = \left(x + \frac{1}{n}, \infty\right),$$

but not by finitely many of these. Thus, K contains all its limit points, and hence K is closed.   □

Typical examples of compact sets are the closed intervals $[a, b]$ in $\mathbb{R}$. Indeed, the nested interval property of Sect. 2.6 has the following generalization to compact sets.

Nested compact sets. *If $K_1 \supseteq K_2 \supseteq K_3 \supseteq \cdots$ are compact sets, then $K_1, K_2, K_3, \ldots$ have a point in common (and the point is unique if the size of K_n tends to zero).*

Proof. The sets $\mathbb{R} - K_n$ are complements of closed sets, hence open. Their union $\bigcup_n (\mathbb{R} - K_n)$ covers the complement of $\bigcap_n K_n$, so if $\bigcap_n K_n$ is empty then $\bigcup_n (\mathbb{R} - K_n)$ covers $\mathbb{R}$. In particular, it covers the interval that contains K_1, which we can take to be $[0, 1]$, without loss of generality.

Thus, we have a covering of $[0, 1]$ by the open sets $\mathbb{R} - K_n$. By compactness, $[0, 1]$ is also covered by finitely many of them, which we can assume to be $\mathbb{R} - K_1, \mathbb{R} - K_2, \ldots, \mathbb{R} - K_m$. But

$$\mathbb{R} - K_1 \subseteq \mathbb{R} - K_2 \subseteq \cdots \subseteq \mathbb{R} - K_m \quad \text{because} \quad K_1 \supseteq K_2 \supseteq \cdots \supseteq K_m,$$

so $K_m \subseteq [0, 1]$ is not covered, which is a contradiction.

Our assumption that $\bigcap_n K_n$ is empty is therefore false; there are points in $\bigcap_n K_n$, and clearly just one point if the size of K_n tends to zero.   □

Exercises

Compact sets are generally "better behaved" than sets that are merely closed, as the nested compact sets property shows.

5.4.1 Give an example to show this is not generally true for closed sets.

5.4.2 Prove that a continuous f maps a compact K onto a compact K'.

5.4.3 Give an example to show that this is not generally true for closed sets.

The following exercises give a proof of the Bolzano–Weierstrass theorem (Sect. 3.6) using the compactness of $[0,1]$ instead of the bisection argument.

5.4.4 Suppose that $S \subset [0, 1]$ is infinite but has no limit point in $[0,1]$. Deduce that each $x \in [0, 1]$ lies in an open interval with no points of S other than (possibly) itself.

5.4.5 From the covering of $[0,1]$ given by Exercise 5.4.4, derive a contradiction by compactness.

5.5 Perfect Sets

The concept of a perfect set goes hand-in-hand with the concept of an isolated point, as we see from the following:

Definition. A point P of a closed set F is *isolated* if there is an ε-neighborhood of P containing no other points of F. A nonempty closed set with no isolated points is said to be *perfect*.

For example, $\mathbb{N}$ is a closed set consisting entirely of isolated points, whereas $[0, 1]$ is a closed set with no isolated points. Other examples of closed sets without isolated points are $\mathbb{R}$ and the Cantor set of Sect. 3.7. We have seen that the last three have continuum cardinality, and in fact Cantor proved that this is true of all perfect sets.

Cardinality of Perfect Sets. *Every perfect set has continuum cardinality.*

Proof. The bijection from $\mathbb{R}$ to $(0, 1)$ introduced in Sect. 3.3 clearly sends perfect sets to perfect sets (no isolated points in the image), hence it suffices to find the cardinality of a perfect set $F \subseteq (0, 1)$. We do this by imitating the proof from Sect. 3.7 that the Cantor set has continuum cardinality.

We construct an infinite binary tree of sets as shown in Fig. 5.3, where each set is perfect and the two sets $F_{\alpha 0}$ and $F_{\alpha 1}$ immediately below each set F_α in the tree are subsets of F_α that we can view as its "lower third" and "upper third."

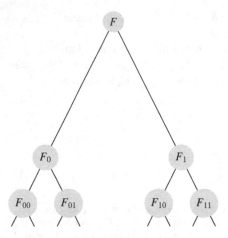

Fig. 5.3 The tree of perfect sets

These "thirds" are found as follows for F, and it is similar for other sets in the tree. First note that F has a minimum member x_0 and a maximum member x_1. Namely,

$$x_0 = \text{lub of } x \text{ for open intervals } (-1, x) \text{ in the complement of } F,$$

and the existence of x_1 is shown similarly. Now consider the set

$$F \cap [x_0, x_0 + (x_1 - x_0)/3].$$

It is the intersection of closed sets, hence closed, and it has no isolated points *except possibly* the upper endpoint $x_0 + (x_1 - x_0)/3$. If this point is in F and isolated, remove it, and in any case call the resulting perfect set F_0 the "lower third of F."

We similarly construct a perfect set F_1, the "upper third of F," by removing the one possible isolated point (if it *is* isolated) from the lower end of the closed set

$$F \cap [x_1 - (x_1 - x_0)/3, x_1].$$

We can then repeat the construction, finding perfect sets F_{00} and F_{01} that are "lower third" and "upper third" of F_0, and perfect sets F_{10} and F_{11} that are "lower third" and "upper third" of F_1; and so on, obtaining the tree of perfect sets shown in Fig. 5.3.

Moreover, the length of the intervals housing the sets tends to zero as one moves down any branch of the tree—each interval being at most 1/3 of the one before—so there is exactly one point common to all the sets on a branch, by the nested compact sets property from the previous section. This point is in F, since all sets in the tree are subsets of F. Finally, the points belonging to different branches are different, since they belong to disjoint intervals. So there are continuum-many points in F, because there are continuum-many branches in the infinite binary tree.   □

It follows from this theorem that any set of real numbers containing a perfect set has the cardinality of the continuum. So if every set S of real numbers has the *perfect set property*—if uncountable, S contains a perfect set—the continuum hypothesis will follow. Indeed, Cantor set out to prove the continuum hypothesis by proving the perfect set property for larger and larger classes of sets. In Chap. 6 we will discuss the first step in his program: the so-called *Cantor–Bendixson theorem* that says any uncountable *closed* set contains a perfect set.

5.5.1 Beyond Open and Closed Sets

There are many sets in $\mathbb{R}^n$ that are neither open nor closed. For example, the set $\mathbb{Q}$ of rational numbers is neither open nor closed in $\mathbb{R}$. It is not open because it does not contain ε-neighborhoods of its members, and it is not closed because it does not contain some of its limit points (e.g., $\sqrt{2}$). Since $\mathbb{Q}$ is an important set, it is desirable to find a classification of sets that goes beyond the open and closed sets. Since we are interested in sets likely to arise in analysis, we might expect the limit operation to generate the sets we want in some systematic manner.

However, since we have already used the complement operation, it turns out to be simpler to use a native set operation, *countable union*, which can generate sets very easily in conjunction with the complement operation. For example, $\mathbb{Q}$ is a countable union of closed sets, namely

$$\mathbb{Q} = \bigcup_{i=1}^{\infty} \{r_i\},$$

where $r_1, r_2, r_3, \ldots$ is an enumeration of the rational numbers. Each *singleton set* $\{r_i\}$ is closed because it is the complement of the open set $(-\infty, r_i) \cup (r_i, \infty)$.

Countable union is also a natural operation in the theory of *measure* that we touched on in Sects. 1.7 and 3.5. The measure of a countable union of disjoint sets corresponds to the countable sum of their measures, which are real numbers. It follows, as we will show in Chap. 9, that we can measure any set built from an open set by complementation and countable unions. But we cannot expect to extend the concept of measure much further than this, because only countable sets of real numbers can be summed.

The sets generated from open sets by complementation and countable union are called the *Borel sets*. The complexity of a Borel set may be measured by the "number of operations" required to build it, but this "number" may well be infinite. In the next chapter we will study the appropriate "numbers" (the *ordinal numbers*) for counting the number of steps in infinite processes such as the generation of Borel sets.

Exercises

The "middle third" set constructed in F not only has the same cardinality as the Cantor set C but is also actually homeomorphic to it.

5.5.1 Use the common tree structure to define a bijection f between C and the "middle third" subset of F.

5.5.2 Show that the bijection f is continuous in both directions.

5.6 Open Subsets of the Irrationals

The foundation for the study of Borel sets is the existence of a *universal* open set $\mathcal{U}$—a two-dimensional open set with horizontal sections that are all the one-dimensional sets. For technical reasons, it is easier to construct such a set using the set $\mathcal{N}$ of irrational numbers in $(0,1)$, rather than the whole closed interval $[0,1]$. In this section we carry out the construction of $\mathcal{U}$ in $\mathcal{N}^2$.

We view $\mathcal{N}$ as the set $\mathbb{N}^{\mathbb{N}}$ of all functions $f : \mathbb{N} \to \mathbb{N}$; that is, all infinite sequences $\langle a_1, a_2, a_3, \ldots \rangle$ of positive integers. For each such sequence we have a real number given by the infinite continued fraction

$$\cfrac{1}{a_1 + \cfrac{1}{a_2 + \cfrac{1}{a_3 + \cfrac{1}{\ddots}}}}$$

which is between 0 and 1 and irrational because any rational number has a finite continued fraction. (See Sects. 2.7 and 2.8 to refresh your memory of continued fractions.) Conversely, any irrational number in $(0,1)$ has an infinite continued fraction of the above form, and hence corresponds to a sequence $\langle a_1, a_2, a_3, \ldots \rangle$ in $\mathcal{N}$. Thus, we can view $\mathcal{N}$ as the set of all irrational numbers in $(0, 1)$.

We can likewise view the open subsets of $\mathcal{N}$ as its intersections with the open sunsets of $(0, 1)$, which we know are unions of rational intervals. However, there is a more natural way to generate the open subsets of $\mathcal{N}$, from open sets corresponding to the finite sequences $\langle a_1, \ldots, a_k \rangle$. For each such sequence, let

$$G(a_1, \ldots, a_k) = \{f \in \mathcal{N} : f(1) = a_1, \ldots, f(k) = a_k\}.$$

Then $G(a_1, \ldots, a_k)$ corresponds to all the continued fractions of the form

$$
\cfrac{1}{a_1 + \cfrac{1}{a_2 + \cfrac{1}{\ddots + \cfrac{1}{a_k + x}}}}
$$

where x is $1/$(an arbitrary infinite continued fraction) and hence x is an arbitrary irrational number in $(0,1)$. This means that $G(a_1,\ldots,a_k)$ is the intersection of N with the open interval in $(0,1)$ whose endpoints are the rational numbers obtained by substituting $x = 0$ and $x = 1$ in the continued fraction above.

It is clear that any $\langle a_1, a_2, a_3, \ldots \rangle \in N$ belongs to all the open sets $G(a_1)$, $G(a_1, a_2)$, $G(a_1, a_2, a_3)$, ... and the size of $G(a_1, \ldots, a_k)$ tends to zero as k increases, by the convergence of continued fractions proved in Sect. 2.8. So, if O is an open subset of N and if $\langle a_1, a_2, a_3, \ldots \rangle \in O$ then $G(a_1, \ldots, a_k) \subseteq O$ for k sufficiently large. It follows that any open set $O \subseteq N$ is a union of sets of the form $G(a_1, \ldots, a_k)$. For this reason, the open sets $G(a_1, \ldots, a_k)$ are called *basic*.

5.6.1   Encoding Open Subsets of N by Elements of N

We know from example 8 in Sect. 3.1 that there is an enumeration of all finite sequences $\langle a_1, \ldots, a_k \rangle$ of natural numbers. If $\langle a_1, \ldots, a_k \rangle$ is the nth sequence in some fixed enumeration, we let $G_n = G(a_1, \ldots, a_k)$. Then any open set $O \subseteq N$ is a union of certain G_n, and hence

$$
O = \bigcup_{n=1}^{\infty} G_{f(n)} \quad \text{for some} \quad f : \mathbb{N} \to \mathbb{N}.
$$

In this way, each $f \in N$ encodes an open subset of N, and if we imagine the elements of N as irrationals on the y-axis, we can envisage the set encoded by y displayed on the horizontal line at height y. Remarkably, the subset of the plane defined in this way is an open subset of $N^2 = N \times N$. We call it a *universal* open set.

Universal open set.   *There is an open subset $\mathcal{U}$ of N^2 whose sections*

$$
\mathcal{U}(y) = \{x : \langle x, y \rangle \in \mathcal{U}\}
$$

are all the open subsets of N.

Proof.   To define $\mathcal{U}$ we interpret each $y \in N$ as a function $y : N \to N$ via the continued fraction for y. Then let

$$
\langle x, y \rangle \in \mathcal{U} \Leftrightarrow x \in G_{y(n)} \quad \text{for some } n.
$$

Thus,

$$\mathcal{U}(y) = \{x : x \in G_{y(n)} \text{ for some } n\} = \bigcup_{n=1}^{\infty} G_{y(n)}$$

is the open subset of $\mathcal{N}$ encoded by y, as described above. As y runs through all the elements of $\mathcal{N}$, the unions $\bigcup_{n=1}^{\infty} G_{y(n)}$ run through all the open subsets of $\mathcal{N}$.

To see why $\mathcal{U}$ is an open subset of $\mathcal{N}^2$, observe that

$$\mathcal{U} = \bigcup_{n=1}^{\infty} H_n, \quad \text{where} \quad H_n = \{\langle x, y \rangle : x \in G_{y(n)}\}.$$

Since $\mathcal{U}$ is the union of the sets H_n, it suffices to prove that each H_n is open. We do this by showing

$$\langle x_0, y_0 \rangle \in H_n \Rightarrow \langle x, y \rangle \in H_n \text{ for all } \langle x, y \rangle \text{ sufficiently close to } \langle x_0, y_0 \rangle.$$

Well, if y is sufficiently close to y_0, the continued fractions for y and y_0 agree to a given depth n, so y and y_0 agree as functions up to a given argument n, which means $G_{y(n)} = G_{y_0(n)}$ for y sufficiently close to y_0.

Also, since $G_{y(n)} = G_{y_0(n)}$ is an open set,

$$x_0 \in G_{y_0(n)} \Rightarrow x \in G_{y_0(n)} \text{ for } x \text{ sufficiently close to } x_0.$$

Putting these two facts together, we get

$$\langle x_0, y_0 \rangle \in H_n \Rightarrow x_0 \in G_{y_0(n)}$$

$$\Rightarrow x \in G_{y(n)} \text{ for } \langle x, y \rangle \text{ sufficiently close to } \langle x_0, y_0 \rangle$$

$$\Rightarrow \langle x, y \rangle \in H_n \text{ for } \langle x, y \rangle \text{ sufficiently close to } \langle x_0, y_0 \rangle,$$

as required. $\square$

Exercises

The above result may be used to construct a universal open set in $[0, 1] \times [0, 1]$.

5.6.1 Show that the closed set $\mathcal{F} = \mathcal{N}^2 - \mathcal{U} \subset \mathcal{N}^2$ is universal in the sense that its sections

$$\mathcal{F}(y) = \{x : \langle x, y \rangle \in \mathcal{F}\}$$

are all the closed subsets of $\mathcal{N}$.

5.6.2 Now consider the *closure* $\overline{\mathcal{F}}$ of $\mathcal{F}$ in $[0, 1] \times [0, 1]$, obtained by adding all the limit points of $\mathcal{F}$ in $[0, 1] \times [0, 1]$. Show that the sections

Fig. 5.4 Felix Hausdorff

$$\overline{\mathcal{F}}(y) = \{x : \langle x, y \rangle \in \overline{\mathcal{F}}\}$$

are all the closed subsets of [0,1], and hence that $\overline{\overline{\mathcal{F}}}$ is a universal closed set for [0,1].
5.6.3 Deduce from Exercise 5.6.2 that $[0, 1] \times [0, 1] - \overline{\mathcal{F}}$ is a universal open set.

5.7 Historical Remarks

According to Ferreirós (1999), p. 139, Dedekind around 1870 developed the ideas of open and closed sets in analysis, but did not publish them. Later they were rediscovered by Peano (1887) and Jordan (1893). Cantor (1884) took up the study of closed sets, as the first stage in his program to prove the continuum hypothesis by showing that each uncountable set of reals contains a perfect subset, and hence has continuum cardinality. He took no interest in open sets, presumably because their cardinality is obvious.

The idea of characterizing continuous functions f as those such that f^{-1} of any open set is open is due to Hausdorff (1914). In the same book, Hausdorff also introduced the concept of a topological space, as one with a system of "open sets" with the three characteristic properties listed in Sect. 5.2. Actually, Hausdorff included a fourth property, stating that any two points lie in disjoint open sets. Topological spaces with the fourth property are now called *Hausdorff spaces*.

Hausdorff also investigated the Borel sets, which arise from the open sets by the operations of complement and countable union. In Hausdorff (1916) he showed that the Borel sets have the perfect set property, thus carrying Cantor's program to a much higher level. The same result was proved independently by Alexandrov (1916). Cantor's program has since been pushed further, but not to all subsets of $\mathbb{R}$. Nevertheless, it remains viable in the sense that it is *consistent* with the usual axioms of set theory for each uncountable set of reals to have a perfect subset. Just what the "usual axioms" are will be explained in the next chapter. Suffice to say, at this point, that there is a "model" of the set theory axioms in which every uncountable set of reals has a perfect subset. The model is due to Solovay (1970), and we say more about it in Sect. 6.8.

Chapter 6
Ordinals

PREVIEW

To formalize the idea of "counting past infinity," we first need a clear idea of the numbers $0, 1, 2, 3, \ldots$ involved in ordinary counting. It appears that the *set* concept is the simplest idea that can serve as a foundation for counting through, and beyond, the finite numbers. The sets involved in the infinite counting process are called *ordinal numbers* or simply *ordinals*, and the natural numbers are represented by sets called the *finite ordinals*.

The finite ordinals can be defined with almost ridiculous ease as follows, using just the concepts of set and membership:

$$0 = \text{empty set},$$

$$1 = \{0\} \quad \text{(the set with member 0)},$$

$$2 = \{0, 1\} \quad \text{(the set with members 0 and 1)},$$

and so on. With this definition, each finite ordinal is the set of all its predecessors, and $m < n$ if and only if $m \in n$. Thus, the $<$ relation is simply membership. It is then natural to take the first infinite ordinal to be

$$\omega = \{0, 1, 2, 3, \ldots\},$$

since its members are precisely the finite ordinals.

This is the right idea, but it involves the assumption that infinite sets exist. Further assumptions about sets are required to push the idea further, to ordinal numbers that are not merely infinite but uncountable. In fact, we end up with a collection of assumptions for sets in general, called the *Zermelo–Fraenkel* (ZF) axioms.

J. Stillwell, *The Real Numbers: An Introduction to Set Theory and Analysis*,
Undergraduate Texts in Mathematics, DOI 10.1007/978-3-319-01577-4_6,
© Springer International Publishing Switzerland 2013

6.1   Counting Past Infinity

Recall from Sect. 5.5 that an *isolated* point of a closed set F is a point $P \in F$ with an ε-neighborhood containing no other points of F, and that a closed set F is *perfect* if it contains no isolated points. These concepts suggest the possibility of finding a perfect subset (if it exists) of a closed set F by repeatedly removing all isolated points. Before we state a theorem to this effect (the *Cantor–Bendixson* theorem of Sect. 6.4), it is well to be aware of what can happen when we "repeatedly" remove isolated points from a closed set.

Cantor (1872) called the result of removing the isolated points from a closed set F the *derived set* F' of F, and he noticed that there are closed sets on which the derived set operation $'$ may be repeated many times. In fact, the operation $'$ may be repeated an infinite number of times, and more. That is, after an infinite sequence of applications of the operation $'$, some members of F can remain, so $'$ can be applied again, perhaps infinitely often. To cope with this unprecedented situation Cantor developed set theory, and particularly the theory of *ordinal numbers*, in order to describe situations in which it is natural to count "past the finite numbers" or *transfinitely*.

Granted that the sequence of operations $'$ may be infinitely long, one still feels that it must eventually be completed, leaving a closed set $F^{(\alpha)}$ with no isolated points. The problem is to describe the number α of times the operation $'$ must be applied and, hopefully, to show that α is *countable* (so as to show that the set of points removed from F is countable).

To give an idea why the $'$ operation may be applied many times, we show how to build more and more complicated closed sets $F \subseteq \mathbb{R}$. We start with

$$F_1 = \left\{ \frac{1}{2}, \frac{3}{4}, \frac{7}{8}, \frac{15}{16}, \ldots, 1 \right\}.$$

F_1 is rather easy to visualize, but we also include a picture (Fig. 6.1) in order to introduce a graphic device that will be useful for more complicated sets. Each point of F_1 lies in the middle of a vertical line—which is easier to see than the point itself—and we make the lines shorter as the points get closer together.

The operation $'$ can be applied exactly twice to F_1, because each of its points except 1 is isolated. But if we make each isolated point of F_1 the limit point of a new sequence of isolated points, as in the set F_2 shown in Fig. 6.2, then $F_2' = F_1$, so the $'$ operation can be applied to F_2 exactly three times.

Fig. 6.1 The set F_1 with derived set $\{1\}$

Fig. 6.2   The set F_2 with derived set F_1

Similarly, we can make each isolated point of F_2 the limit point of a sequence of isolated points of a set F_3, so that the operation $'$ applies exactly four times to F_3, and so on. For each natural number n we can construct a closed set F_n to which the operation $'$ applies exactly $n + 1$ times.

This is only the beginning. We now construct a set F_ω to which the operation $'$ applies *infinitely often*, by arranging that, in F_ω,

$$\frac{1}{2} \text{ is the limit of a set like } F_1 \text{ lying between 0 and } \frac{1}{2},$$

$$\frac{3}{4} \text{ is the limit of a set like } F_2 \text{ lying between } \frac{1}{2} \text{ and } \frac{3}{4},$$

$$\frac{7}{8} \text{ is the limit of a set like } F_3 \text{ lying between } \frac{3}{4} \text{ and } \frac{7}{8},$$

$$\vdots$$

This can obviously be done by suitably scaling and translating the sets $F_1, F_2, F_3, \ldots$. Then

$$\text{One application of } ' \text{ removes all points in } F_\omega \text{ that are } < \frac{1}{2}$$

$$\text{Two applications of } ' \text{ remove all points in } F_\omega \text{ that are } < \frac{3}{4}$$

$$\text{Three applications of } ' \text{ remove all points in } F_\omega \text{ that are } < \frac{7}{8}$$

$$\vdots$$

so we can say that the operation $'$ applies to F_ω *as many times as there are natural numbers*. This number of applications is denoted by ω (and this is why we have already used ω for the subscript of the corresponding closed set). And, in fact to remove the limit point 1 of F_ω we need to apply the operation $'$ *once more* after as many steps as there are natural numbers. Naturally, we denote this number of applications by $\omega + 1$.

The numbers ω and $\omega + 1$ are the first members of what Cantor called the *second number class*. Today, we call these numbers (infinite) *countable ordinals*. (The *first number class* consists of the finite ordinals $0, 1, 2, \ldots$.)

Exercises

Several operations we have already seen in this book invite "transfinite continuation" like that we have just seen for the ' operation.

6.1.1 *Diagonalization of real numbers.* Given real numbers $x_1, x_2, x_3, \ldots$, we can use the diagonal argument to get a new number, which we might call x_ω. Then we can also diagonalize the sequence $x_\omega, x_1, x_2, x_3, \ldots$ to get $x_{\omega+1}$. See how far you can continue.

6.1.2 *Diagonalization of integer functions.* Suppose we have a sequence of increasing functions $f_i : \mathbb{N} \to \mathbb{N}$, each of which grows faster than the one before. That is $f_{i+1}(n)/f_i(n) \to \infty$ as $n \to \infty$. Define a function f_ω that grows faster than each f_i.

6.1.3 *Classes of functions obtained as limits.* Let $B_0 =$ {continuous real functions} and $B_{n+1} =$ {limits of functions in B_n}. Supposing that each B_{n+1} has members not in B_n, how might B_ω be defined?

6.2   What Are Ordinals?

In the previous section and its exercises we have seen several kinds of operation that can be applied infinitely often. The sequence of applications may be not merely infinite, but "longer than the sequence of natural numbers." We can introduce symbols $\omega, \omega + 1, \omega + 2, \ldots$ for the numbers ("ordinals") that measure the length of these infinite sequences, but the meaning of the symbols will remain hazy until we have a precise definition of what ordinals are. Cantor used set theory in an informal way to study ordinals, but von Neumann (1923) was the first to provide a definition of ordinal numbers as a particular kind of set. Von Neumann's definition is so elegant that it throws new light on the natural numbers—the *finite* ordinals—as well.

6.2.1   Finite Ordinals

The number 0 is defined to be the *empty set*, denoted by { } or $\emptyset$. Then $1, 2, 3, \ldots$ are defined successively as follows:

$$0 = \{\ \},$$
$$1 = \{0\},$$
$$2 = \{0, 1\},$$
$$3 = \{0, 1, 2\},$$
$$\vdots$$

In a nutshell, $n + 1$ is the set with members $0, 1, 2, \ldots, n$, so *each nonzero natural number is the set of its predecessors.* This magical definition, which seems to make numbers out of nothing, also connects the native concepts of set theory—membership and union—with the native concepts of number theory—order and successor. Indeed, we have:

- For any natural numbers m and n, $m < n \Leftrightarrow m \in n$.
- The successor function $S(n) = n \cup \{n\}$.

As we saw in Sect. 2.2, Grassmann discovered that number theory can be based on the successor function, by using successor to define addition and multiplication. By combining Grassmann's conception of number theory with von Neumann's concept of natural number, we see that number theory is part of set theory. In fact, Ackermann (1937) showed that number theory is essentially identical to finite set theory. We will return to this surprising view of number theory in Sect. 6.6, when we have developed a clearer picture of what set theory actually is.

Our immediate task is to explain the concept of *infinite* ordinals, which will force us to examine the set concept more closely.

6.2.2   Infinite Ordinals: Successor and Least Upper Bound

To define infinite ordinals, we develop one of the key ideas of the previous section: *an ordinal is the set of all its predecessors*. Given that the infinite set

$$\omega = \{0, 1, 2, 3, \ldots\}$$

of all natural numbers exists, we can say that ω is the least infinite ordinal because its predecessors (i.e., members) are all the finite ordinals. We can also say that ω is the *least upper bound* of the finite ordinals, because it is greater than them all, but no smaller ordinal is. Thus, the step from finite to infinite ordinals demands the existence of the infinite set ω. This set is in fact the foundation of all infinite set theory, but for now we will be content to show how ω is the foundation of the infinite ordinal numbers.

The successor operation $S(x) = x \cup \{x\}$ applies to any set. So, starting with ω, we can generate an infinite sequence of infinite ordinals:

$$\omega + 1 = \{0, 1, 2, 3, \ldots, \omega\},$$

$$\omega + 2 = \{0, 1, 2, \ldots, \omega, \omega + 1\},$$

$$\omega + 3 = \{0, 1, 2, \ldots, \omega, \omega + 1, \omega + 2\},$$

$$\vdots$$

Then, by embracing[1] all the sets created so far in a single set, we obtain their least upper bound

---

[1] Happily, set theory literally uses the braces { and } to comprehend a collection of objects as a set.

$$\omega \cdot 2 = \{0, 1, 2, \ldots, \omega, \omega + 1, \omega + 2, \ldots\}.$$

Another sequence of successors then leads to

$$\omega \cdot 3 = \{0, 1, 2, \ldots, \omega, \omega + 1, \omega + 2, \ldots, \omega \cdot 2, \omega \cdot 2 + 1, \omega \cdot 2 + 2, \ldots\},$$

and we similarly obtain $\omega \cdot 4, \omega \cdot 5, \ldots$.

The sequence of ordinals $\omega, \omega \cdot 2, \omega \cdot 3, \ldots$ also has a least upper bound, which we obtain by collecting all of these sets, and all their predecessors, into a single set ω^2. Since predecessors are members, this least upper bound is simply the *union* of the sets $\omega, \omega \cdot 2, \omega \cdot 3, \ldots$. That is

$$\omega^2 = \omega \cup \omega \cdot 2 \cup \omega \cdot 3 \cup \cdots.$$

As we ascend to larger ordinals, it becomes more and more convenient to take unions of infinitely many sets to obtain least upper bounds. Indeed, the union of any set of ordinals is the least upper bound of the set.

In this way we can grasp ordinals $\omega^3, \omega^4, \omega^5, \ldots$ and their least upper bound ω^ω; then ordinals $\omega^{\omega^2}, \omega^{\omega^3}, \omega^{\omega^4}, \ldots$ and their least upper bound ω^{ω^ω}; and so on, to ever more dizzying heights.

Yet, mind-boggling as these ordinals may be, they are all *countable*. That is, they are sets with countably many members, because every least upper bound operation applied so far involves a countable union of countable sets.

6.2.3 Uncountable Ordinals

Most of the ordinals encountered in this book are countable, but it should be no surprise that uncountable ordinals exist. In fact, the *least* uncountable ordinal, ω_1 should be the set of all countable ordinals. However, to make this definition of ω_1 precise we need a precise definition of ordinal and $<$. Here it is.

Definition. An *ordinal* α is a set that is

- $\in$-transitive: that is, if $\beta \in \alpha$ and $\gamma \in \beta$, then $\gamma \in \alpha$,
- $\in$-linear: that is, if $\beta, \gamma \in \alpha$, then either $\beta \in \gamma$ or $\gamma \in \beta$.

Also $\beta < \alpha$ if and only if $\beta \in \alpha$.

It is not hard to check that this definition is satisfied by all the ordinals mentioned so far. However, the nature of the sets that satisfy the definition depends on the nature of the membership relation $\in$, and hence ultimately on the axioms of set theory. We discuss these axioms more fully later, but one should be mentioned here because it is motivated by the properties of ordinals. This is the *axiom of foundation*, which says that there is no infinite descending membership sequence $\cdots \in \alpha_3 \in \alpha_2 \in \alpha_1$. Among other things, the axiom of foundation ensures that each set of ordinals has a least member, and hence it enables definitions and proofs by *induction* on ordinals. This extended form of induction is called *transfinite induction*.

Exercises

Use the definition of ordinal to verify the following.

6.2.1 If β is an ordinal and $\alpha \in \beta$, then α is an ordinal.
6.2.2 If α and β are ordinals and $\alpha \subseteq \beta$, then $\alpha \leq \beta$.
6.2.3 If α is an ordinal, then so is $\alpha + 1 = \alpha \cup \{\alpha\}$.
6.2.4 If $\alpha_1 < \alpha_2 < \cdots$ are ordinals, so is their lub, $\bigcup_i \alpha_i$.
6.2.5 Also verify that $\bigcup_i \alpha$ is indeed the lub of the α_i, because any ordinal less than $\bigcup_i \alpha$ is less than some α_i.

6.3 Well-Ordering and Transfinite Induction

The usual way to state the axiom of foundation, which obviously implies the nonexistence of infinite descending membership sequences, is the following:

Axiom of Foundation. Each nonempty set S has an $\in$-least member; that is, an element $x \in S$ such that $y \notin x$ for any $y \in S$.

It follows, in particular, that $x \notin x$ for any set x, and the definition of ordinal implies that any ordinal σ is *linearly ordered*[2] by the membership relation $\in$. That is, if we write the usual order symbol $<$ in place of $\in$, then any α, β, γ in σ satisfy:

1. $\alpha \not< \alpha$ (Irreflexivity)
2. If $\alpha \neq \beta$ then either $\alpha < \beta$ or $\beta < \alpha$ (but not both).   (Linearity)
3. If $\alpha < \beta$ and $\beta < \gamma$ then $\alpha < \gamma$.   (Transitivity)

The axiom of foundation gives a fourth property that makes the linear ordering a *well*-ordering:

4. Any subset of σ has least member.   (Well-foundedness)

It follows that σ has a least member, which can only be the empty set 0. Because if the least $\alpha \in \sigma$ were not empty, any $\beta \in \alpha$ would be a lesser member of σ, contradicting the definition of α. It similarly follows that the least member of $\sigma - \{0\}$ is $\{0\} = 1$, and so on. Indeed, it is easy to see that σ is the set of all ordinals less than σ.

---

[2]We wrote down the defining properties of a linear ordering once before, in Sect. 2.5. There we stated them as properties of $\leq$, because they were motivated by the $\subseteq$ relation between lower Dedekind cuts. Here we are motivated by the $\in$ relation, so we write them as properties of $<$. However, it is easy to see that the two sets of properties are equivalent.

Ordinals are not the only examples of well-ordered sets, but they include the *order types* of all well-ordered sets, in the following sense.

Ordinal Representation of Well-orderings. *If $<$ is a well-ordering relation on a set W, then $\langle W, < \rangle$ is order-isomorphic to some ordinal σ under the $\in$ relation. That is, there is a bijection $f : \sigma \to W$ such that*

$$\alpha \in \beta \Leftrightarrow f(\alpha) < f(\beta).$$

Proof. We would like to define f "inductively" by saying

$$f(0) = \text{ least member of } W,$$

$$f(\alpha) = \text{ least member of } W - \{f(\beta) : \beta < \alpha\},$$

until we reach an ordinal σ such that $\{f(\beta) : \beta < \sigma\} = W$. However, we have not yet proved that such a *transfinite induction* is valid, so we take the following more cautious approach.

Consider all the bijections f_α (between ordinals α and subsets of W) satisfying the following conditions:

1. $f_\alpha(\beta)$ is defined for all $\beta < \alpha$.
2. $f_\alpha(0) = $ least member of W.
3. For any $\gamma < \alpha$, $f_\alpha(\gamma) = $ least member of $W - \{f_\alpha(\beta) : \beta < \gamma\}$.

The set of such functions is not empty, since it includes the function f_1 consisting of the single ordered pair $\langle 0, \text{ least member of } W \rangle$.

Also, any two such functions, say f_α and g_δ, are *compatible* in the sense that $f_\alpha(\gamma) = g_\delta(\gamma)$ on any γ on which they are both defined. Because if $f_\alpha(\gamma) \neq g_\delta(\gamma)$ then there is a *least* γ for which this happens, and one sees that this least γ contradicts conditions 2 and 3 above. Compatibility implies that, for each α, there is *at most one* function f_α satisfying conditions 1, 2, and 3.

Now let σ be the least ordinal greater than all the α for which f_α exists. By compatibility, the union f of $\{f_\alpha : \alpha < \sigma\}$ is an injection $f : \sigma \to W$. If f is not onto W then $W - \{f(\alpha) : \alpha < \sigma\}$ is not empty, and we can define

$$f(\sigma) = \text{ least member of } W - \{f(\alpha) : \alpha < \sigma\},$$

which contradicts the definition of σ.

Thus, f is a bijection $\sigma \to W$, and it easily follows from conditions 2 and 3 that f is order-preserving. (If not, take the least α and β such that $\alpha \in \beta$ but $f(\alpha) > f(\beta)$, and derive a contradiction.) $\square$

This theorem shows why any well-ordered set is isomorphic to an ordinal. The proof also shows how one may justify defining a function by transfinite induction— that is, by taking the union of all functions that satisfy the induction up to a certain ordinal. From now on we will use inductive definitions without detailed justification.

Another important corollary of this proof is the following.

Corollary 3. *For any two ordinals μ and ν, either $\mu \in \nu$ or $\nu \in \mu$.*

Proof. As above, take the union of all bijections f_α from some $\alpha < \mu$ to $\alpha < \nu$. The set is not empty because it includes the function f_1 consisting of the single ordered pair $\langle 0, 0 \rangle$. If we now take σ to be the least ordinal greater than the α for which f_α exists, then we find $\sigma = \mu$ or $\sigma = \nu$. It follows that $\mu \in \nu$ or $\nu \in \mu$. $\qquad\square$

Exercises

In Sect. 6.1 we gave an example of a well-ordered set of rationals with the order type ω; namely, $\left\{ \frac{1}{2}, \frac{3}{4}, \frac{7}{8}, \ldots \right\}$. We also indicated, by a picture, how to construct a set of numbers with order type ω^2.

6.3.1 Give an explicit set of rational numbers with order type ω^2.

6.3.2 Given a well-ordered set of rationals with order type α, explain how to obtain a set of rational with order type $\alpha + 1$.

6.3.3 Given well-ordered sets of rationals with order types $\alpha_1, \alpha_2, \alpha_3, \ldots$, explain how to construct a set of rationals with order type at least $\bigcup_i \alpha_i$.

6.3.4 Deduce, from Exercises 6.3.2 and 6.3.3 that there are sets of rationals with the order types of all countable ordinals.

The inductive definitions of sum and product from Sect. 2.2 are easily extended to all ordinals by transfinite induction. The "induction step" must now be supplemented by a step for ordinals that are not successors, the so-called *limit* ordinals. Here is the definition of $\alpha + \beta$ by induction on β:

$$\alpha + 0 = \alpha$$

$$\alpha + (\beta + 1) = (\alpha + \beta) + 1$$

$$\alpha + \gamma = \operatorname*{lub}_{\beta < \gamma} (\alpha + \beta) \quad \text{for a limit ordinal } \gamma.$$

6.3.5 Using the definition of sum, prove by induction on γ that the *associative law* holds for ordinal addition: $\alpha + (\beta + \gamma) = (\alpha + \beta) + \gamma$.

6.3.6 Give an example to show that the commutative law does not hold for ordinal addition.

6.3.7 Give an inductive definition of ordinal multiplication, and show that it satisfies the associative law.

6.4 The Cantor–Bendixson Theorem

The first definition by transfinite induction that we will use is that of the sequence of derived sets of a closed set, which we began to discuss in Sect. 6.1. As in the exercises above, we use the term *limit* for an ordinal that is not a successor.

Definition. If F is a closed set, then $F^{(\alpha)}$ is defined for all countable ordinals α as follows:

$$F^{(1)} = F - \{\text{all isolated points of } F\}.$$

$$F^{(\alpha+1)} = F^{(\alpha)} - \{\text{all isolated points of } F^{(\alpha)}\}.$$

$$F^{(\lambda)} = \bigcap_{\alpha<\lambda} F^{(\alpha)} \quad \text{when } \lambda \text{ is a limit ordinal.}$$

In a construction like this one, where points are removed at successor stages $\alpha+1$, taking the intersection is the natural thing to do at a limit stage λ, because $\bigcap_{\alpha<\lambda} F^{(\alpha)}$ omits all the points removed from $F^{(\alpha)}$ at stages $\alpha + 1 < \lambda$.

The above definition makes sense even for uncountable λ, but we are about to show that the sequence $F^{(\alpha)}$ becomes constant at some countable α, so it is pointless to go to uncountable stages. On the other hand, it should be clear from Sect. 6.1 and its exercises that $F^{(\alpha)}$ can continue to change up to an arbitrarily high countable ordinal α. Thus, the theorem on the eventual constancy of $F^{(\alpha)}$—the famous *Cantor–Bendixson theorem*—is a subtle one depending on the general concept of countable ordinal.

The key to the proof of the Cantor–Bendixson theorem is the following theorem, which limits the length of a well-ordered nested sequence of open sets.

Length of a well-ordered nested sequence of open sets. *If we have open sets U_α for $\alpha \leq$ some ordinal γ, and if $\alpha < \beta \leq \gamma \Rightarrow U_\alpha \subsetneq U_\beta$, then γ is countable.*

Proof. Since $U_\alpha \subsetneq U_{\alpha+1}$, for each $\alpha < \gamma$ we have a point $x_\alpha \in U_{\alpha+1} - U_\alpha$ and hence a rational open interval I_α such that $x_\alpha \in I_\alpha \subset U_{\alpha+1}$. Indeed, we can *define I_α* explicitly as the first interval I (in some fixed enumeration of the rational intervals) such that $I \subset U_{\alpha+1}$ but $I \not\subset U_\alpha$.

The intervals I_α, I_β are necessarily different for $\alpha < \beta$, since $I_\alpha \subset U_\beta$ but $I_\beta \not\subset U_\beta$, hence there are only countable many ordinals $< \gamma$, because there are only countably many rational intervals.   □

Cantor-Bendixson Theorem. *If F is a closed subset of $\mathbb{R}$ and $F^{(\alpha)}$ denotes the αth derived set of F, then $F^{(\alpha)} = F^{(\alpha+1)}$ for some countable α, and hence $F^{(\alpha)}$ is either empty or perfect.*

Proof. It follows from the definition of the sets $F^{(\alpha)}$ that they are all closed and that $F_\alpha \supseteq F_\beta$ for $\alpha < \beta$. Hence the open complements U_α of the $F^{(\alpha)}$ are such that $U_\alpha \subseteq U_\beta$ for $\alpha < \beta$.

Then, since a nested well-ordered sequence of open sets has countable length, it follows that $U_\alpha = U_{\alpha+1}$ for some countable ordinal α, and hence $F^{(\alpha)} = F^{(\alpha+1)}$.   □

Exercises

The following exercises explain why we cannot simply find the interval I_α inside $U_{\alpha+1} - U_\alpha$ in the proof of the theorem on the length of a sequence of nested open sets.

6.4.1   Give an example of open sets U, V with $U \subset V$ but with no open interval $I \subset V - U$.

6.4.2   For your example of open sets U, V above, find an open interval I such that $I \subset V$ but $I \not\subset U$.

There is an easier theorem about sequences of *disjoint* open sets.

6.4.3   Show that any collection of disjoint open intervals in $\mathbb{R}$ is countable.

6.4.4   Use Exercise 6.3.4 to construct, for any countable ordinal γ, disjoint half-open intervals $[a_\alpha, a_{\alpha+1})$ for all $\alpha < \gamma$ with the properties

$$a_\alpha < a_\beta \Leftrightarrow \alpha < \beta \quad \text{and} \quad \bigcup_{\alpha < \gamma} [a_\alpha, a_{\alpha+1}) = [0, 1).$$

Since each $[a_\alpha, a_{\alpha+1})$ is homeomorphic to $[0,1)$, it follows that $\bigcup_{\alpha < \gamma} [a_\alpha, a_{\alpha+1})$ is homeomorphic to a structure we may call the γ-*line* $[0, 1) \times \gamma$. The γ-line contains a copy $[0, 1) \times \{\alpha\}$ of $[0,1)$ for each $\alpha < \gamma$, with copy α to the left of copy β when $\alpha < \beta$, and the point 0 in copy β is the least upper bound of all points in the copies α for $\alpha < \beta$.

6.4.5   Deduce from Exercise 6.4.4 that the γ-line is homeomorphic (and order isomorphic) to $[0,1)$ for each countable ordinal γ.

6.4.6   Similarly define the ω_1-*line*, and explain why the ω_1-line is not homeomorphic (or order isomorphic) to $[0,1)$.

The ω_1-line is also known as the *long line*.

6.5   The ZF Axioms for Set Theory

It should now be clear that the set concept is practically indispensable for the study of analysis in general and the real numbers in particular. Moreover, we have seen that even the most basic mathematical objects—the natural numbers—can be naturally defined as certain sets. Certain *axioms* for sets also appear to be indispensable. For example, we need to assume the existence of the empty set and an infinite set (most conveniently, the set of natural numbers). In this section we list the most commonly used axioms for set theory, with comments on their role as a foundation for mathematics. They are called the *Zermelo–Fraenkel axioms*, after Ernst Zermelo who proposed most of them in Zermelo (1908), and Abraham Fraenkel who made an important amendment in Fraenkel (1922). For short, they are called the ZF axioms.

The underlying idea of the ZF axioms is that "everything is a set"; in particular, natural numbers, real numbers, and functions are certain kinds of sets. In line with this conceptual economy, all *relations* between sets are based on *membership* $\in$ and *equality* $=$. Thus, the language of ZF set theory is very simple: it has variables $x, y, z, \ldots$ to denote sets, the relation symbols $\in$ and $=$, and symbols for the basic concepts of logic—"and," "or," "not," "for all," and "there exists." We are not going to use formal logic symbols in this book, but it is necessary to know that they exist, and hence that there is a mathematical precise concept of "formula in the language of ZF set theory." This is because the ZF axioms include an infinite list of formulas, called the *replacement schema*.

Most ZF axioms describe operations for producing new sets from old, by clearly defined processes. Implicitly, they describe the *cumulative concept of set*, according to which each set is constructed from previously defined sets (starting with the empty set). Thus, sets arise in stages, which turn out to be *ordinal number* stages. At no stage does one have a "set of all sets," because there is always a next stage, at which new sets arise. In this way we avoid paradoxes that arose in the early history of set theory.

Extensionality. *Two sets are equal if and only if they have the same members.*
     It follows, for example, that

$$\{0,1\} = \{1,0\} = \{1,1,0\},$$

because each of these sets has the same members, namely, 0 and 1.

Empty Set. *There is a set with no members.*
     It follows from Extensionality that the empty set (which we call 0 from now on) is unique. In the cumulative hierarchy of sets, 0 is at the bottom level.

Pairing. *For any sets X and Y there is a set $\{X, Y\}$ whose only members are X and Y.*
     From the empty set 0 that we have from the previous axiom, we can now construct the set $\{0, 0\}$, which equals $\{0\}$ by Extensionality. Thus, $1 = \{0\}$ occurs at the next level of the cumulative hierarchy.
     By further use of pairing we can construct $2 = \{0, 1\}$, but how do we construct $3 = \{0, 1, 2\}$? See the next axiom.
     The pairing axiom gives us the *unordered* pair $\{X, Y\}$, but there is a trick (due to Kuratowski 1921) which also gives the *ordered* pair $\langle X, Y \rangle$. Namely, let

$$\langle X, Y \rangle = \{\{X\}, \{X, Y\}\}.$$

This definition, clumsy though it may be, has the essential property that

$$\langle X_1, Y_1 \rangle = \langle X_2, Y_2 \rangle \Leftrightarrow X_1 = X_2 \text{ and } Y_1 = Y_2.$$

Union. *For any set X there is a set whose members are the members of members of X.*
     In the case where $X = \{Y, Z\}$, the members of the members of X form what we call the union *of Y and Z, $Y \cup Z$.* This special case of union suffices to form

$$3 = \{0, 1, 2\} = \{0\} \cup \{1, 2\}$$

from sets previously defined by pairing, and more generally we get

$$n + 1 = n \cup \{n\}$$

An important application of union for infinite sets X is where X is a set of ordinals. In this case the set of members of the members α_i of X, $\bigcup_i \alpha_i$, is the least upper bound of the ordinals α_i.

Infinity. *There is an infinite set; in fact a set that includes 0 and, along with any member X, also the member $X \cup \{X\}$.*

Thus, the members of this set include all the finite ordinals. However, we do not yet have the set ω whose members are *exactly* the finite ordinals. To obtain it, we would like an axiom that guarantees the existence of "definable subsets," because we can write down a definition of finite ordinal. Zermelo proposed such an axiom, and Fraenkel proposed something stronger, involving *definable functions*.

Replacement. *For any function definition f, the values $f(x)$, where x is a member of a set X (the* domain *of f), form a set $f(X)$ (the* range *of f).*

Replacement is actually an infinite schema of axioms, one for each two-variable formula $\varphi(x, y)$ written in the language of ZF. Such a formula defines a function f if, for each $x \in X$, $\varphi(x, y)$ holds for exactly one y [called the function value $f(x)$].

The "definable subset" axiom used by Zermelo is the special case of Replacement where f maps the set X onto a subset of itself. For example, if we want to obtain the subset $\omega = \{0, 1, 2, \ldots\}$ from a set Y whose members y include $0, 1, 2, \ldots$ we define f on Y by letting $f(y) = 0$ if y is *not* a finite ordinal, and let $f(y) = y$ otherwise.

Power Set. *For any set X there is a set $\mathcal{P}(X)$ whose members are the subsets of X.*

$\mathcal{P}(X)$ is called the *power set* of X, and we have already seen one way in which Power Set is a "powerful" axiom, in Sect. 3.8. By the diagonal argument, $\mathcal{P}(X)$ is a set of higher cardinality than X. In particular, $\mathcal{P}(\omega)$ is an uncountable set.

The power set axiom is also needed to prove the existence of the least uncountable ordinal ω_1. In fact, *any* proof that uncountable sets exist needs the power set axiom, because the other ZF axioms can be satisfied by a collection of countable sets. (Similarly, Infinity is needed to prove the existence of infinite sets, because the other ZF axioms can be satisfied by a collection of finite sets; see the exercises).

Foundation. *Every set X has a $\in$-minimal member; that is, an $x \in X$ such that $y \in x$ for no $y \in X$.*

As we have already seen, Foundation guarantees that any set that is linearly ordered by the membership relation $\in$ is in fact *well*-ordered by $\in$, which simplifies the definition of ordinal.

Foundation also guarantees the cumulative set concept, by ensuring that each set X has a *rank α*—an ordinal number that counts the number of applications of the power set axiom needed to build X, starting from the empty set. We elaborate on the concept of rank in the exercises and the next two sections.

Exercises

A collection of sets called the *hereditarily finite* sets is obtained by the following inductive construction, already mentioned in the exercises to Sect. 3.8.

$$V_0 = 0, \quad \text{the empty set,}$$

$$V_{n+1} = V_n \cup \mathcal{P}(V_n).$$

The union $V_\omega = \bigcup_n V_n$ is the set of hereditarily finite sets.

6.5.1 Prove by induction on n that $V_{n+1} = \mathcal{P}(V_n)$.

6.5.2 Prove by induction on n that each member of V_n is finite, and hence that members of members, members of members of members, and so on, are all finite.

6.5.3 Explain why V_ω satisfies all ZF axioms except Infinity.

The following exercises explore the idea of a "set of all sets" and the contradictions to which it leads.

6.5.4 If X is the set of all sets, why is $\mathcal{P}(X)$ contradictory?

6.5.5 In particular, what about $Y = \{Z \in X : Z \notin Z\}$?

6.6   Finite Set Theory and Arithmetic

As we saw in Sect. 6.2, the numbers $0, 1, 2, 3, \ldots$ can be taken to be the sets

$$0 = \{\}, \quad 1 = \{0\}, \quad 2 = \{0, 1\}, \quad \ldots,$$

with the successor function $S(n) = n \cup \{n\}$. Thus, ZF can prove the existence of the basic objects of arithmetic. In fact, if we omit Infinity from the ZF axioms, the remaining axiom set ZF−Infinity is equivalent to the Grassmann–Dedekind–Peano axioms mentioned in Sect. 2.2. This is because we do not need Infinity to obtain the individual sets $0, 1, 2, 3, \ldots$ and the successor function, and the axiom of foundation gives us definition and proof by induction.

In a little more detail, here is how the foundations of arithmetic can be established in ZF−Infinity.

1. The natural numbers $0, 1, 2, 3, \ldots$ are the finite ordinals. We gave the definition of "α is an ordinal" in Sect. 6.2. An ordinal α is finite if α and all of its members each have a greatest member, where γ is the greatest member of β if $\gamma \in \beta$ and $\gamma \notin \delta$ for any $\delta \in \beta$.

2. Induction amounts to the principle that, if some natural number n has property P, then there is a least natural number with property P (for properties P definable in the language of ZF). Since n is a finite ordinal, the numbers $\leq n$ are the members of $S(n) = n \cup \{n\}$, so the least number with property P is the least member of the set

$$\{m : m \in n \cup \{n\} \text{ and } m \text{ has property } P\}.$$

This least member exists by the foundation axiom.

3. Since induction is available, we can define sum and product by induction, as in Sect. 2.2. All other functions and theorems of arithmetic are then obtainable by induction.

Of course, the subject matter of ZF−Infinity is more than just the finite ordinals $0, 1, 2, 3, \ldots$—but not *much* more, as it happens. The only sets that ZF−Infinity can prove to exist are those obtained from the empty set 0 by iterating the power set operation a finite number of times. (These are the members of the sets V_n discussed in the exercises to the previous section, where it was shown that they satisfy all the axioms of ZF−Infinity. Consequently, the sentence "every set belongs to some V_n" is consistent with ZF−Infinity.)

These finite sets can be encoded by natural numbers, and set operations such as pairing and union can then be interpreted as operations on numbers. This "arithmetization" of finite set theory is based on the ideas of Gödel (1931), who arithmetized formal logic to prove his famous theorem on the incompleteness of formal systems for arithmetic. The details of arithmetization are tedious and do not concern us, but it is useful to know that arithmetic is strong enough to interpret other systems for operating on finite objects. Its ability to interpret finite set theory is the reason we say that arithmetic is equivalent to ZF−Infinity.

Exercises

A typical example of arithmetization is the encoding of an ordered pair $\langle m, n \rangle$ of natural numbers by a single natural number.

6.6.1 Give ways to encode $\langle m, n \rangle$ by a natural number; (i) using only addition and multiplication, and (ii) using exponentiation.

6.6.2 Also give an inductive definition of exponentiation, assuming definitions of addition and multiplication.

Ordered pairs are useful for extending the arithmetic of natural numbers to integers and rational numbers.

6.6.3 Suppose we want the pair $\langle a, b \rangle$ to behave like $a - b$. Under what conditions do $\langle a, b \rangle$ and $\langle a', b' \rangle$ represent the same integer?

6.6.4 Also define addition and multiplication of pairs so as to reflect addition and multiplication of integers.

6.6.5 Suppose we want the pair $\langle a, b \rangle$ to behave like a/b. Under what conditions do $\langle a, b \rangle$ and $\langle a', b' \rangle$ represent the same rational number?

6.6.6 Also define addition and multiplication of pairs so as to reflect addition and multiplication of rationals.

6.7 The Rank Hierarchy

The claim that ZF captures the idea of building the universe of sets in ordinal-numbered stages is formalized by a hierarchy of sets V_α, and the associated concept of *rank*, which are defined by an induction of all the ordinal numbers α.

Definition. Sets V_α are defined as follows:

- $V_0 = 0$ (the empty set),
- $V_{\alpha+1} = \mathcal{P}(V_\alpha)$,
- $V_\lambda = \bigcup_{\beta < \lambda} V_\beta$ for each limit ordinal λ.

The *rank* of a set X is defined inductively as the least α such that each member of X has rank less than α.

Thus, V_α may be viewed as the set of all sets built using $< \alpha$ applications of the power set operation $\mathcal{P}$, and the claim that every set is built at some stage is:

Existence of Rank. *Each set X has a rank.*

Proof. Suppose X is a set that does *not* have a rank. Then some member X_1 of X also does not have a rank. Because if each $x \in X$ has a rank, $\text{rank}(x)$, the replacement axiom tells us that the range of the rank function on X is a set of ordinals, with union α say. This means that X has a rank ($\leq \alpha + 1$), contrary to assumption.

Thus, there are members X_1 with no rank, and similarly members X_2 of these X_1 with no rank, and so on. With the help of the replacement and union axioms we can collect these

members X_1 of X with no rank,

members X_2 of members of X with no rank,

. . . .

into a single set N. But then N is a set with no $\in$-minimal member, contrary to the foundation axiom.   □

The universe of all sets can therefore be viewed as the union of the sets V_α, as α ranges over all the ordinals. It is natural to use the symbol V to denote the universe of all sets, though one should always remember that V *is not a set*. [If it were, $\mathcal{P}(V)$ would have cardinality greater than the universe, which is absurd.]

6.7.1   Cardinality

In Sect. 3.3 we defined sets to be of the *same cardinality* if there is a bijection between them. This suggests that such sets share a common property, called *cardinality* or *cardinal number*, which we have not yet defined. Up until now, the problem in defining "cardinality" is that the collection of all sets with the same cardinality is not a set (for much the same reason that the union V of all V_α is not a set). But with the help of the concept of rank we can get around this problem as follows.

Definition. For any set X, let α be the minimal rank of a set with the same cardinality as X. Then the *cardinality of X* is the set

$$\{Y \in V_\alpha : Y \text{ has the same cardinality as } X\}.$$

It still does not seem right to call this set a cardinal *number*, because it is not clear that cardinalities can be ordered. It would be simpler if each set X had the same cardinality as an ordinal, in which case we could take the least such ordinal as the cardinal number of X. This can in fact be achieved with an extra axiom, the *axiom of choice*, which is commonly added to ZF for this reason. There are in fact many advantages to the axiom of choice, and some disadvantages, which we discuss in the next chapter.

Exercises

6.7.1 Show that the rank of an ordinal α is α.

6.7.2 Show that the collection of all ordinals is also not a set.

6.7.3 By making suitable definitions of rational numbers and real numbers, find the ranks of $\mathbb{Q}$ and $\mathbb{R}$.

Also locate the following sets in the rank hierarchy.

6.7.4 The set $\mathbb{N} \times \mathbb{N}$ of all ordered pairs from $\mathbb{N}$.

6.7.5 The set of all functions: $\mathbb{N} \to \mathbb{N}$.

6.8 Large Sets

In Sect. 3.8 we claimed that there are "largeness" properties so extreme that sets with those properties cannot be proved to exist. We suggested that one such "largeness" property is *inaccessibility*, where an inaccessible set is one that has infinite members and is closed under the operations of power set and taking ranges of functions. It should now be apparent that *if V_α is an inaccessible set, then V_α satisfies the ZF axioms.*

Certainly, if V_α is large enough to have an infinite member, then it satisfies the empty set and infinity axioms. It satisfies power set and replacement by the hypothesis of closure under power set and taking ranges of functions. Closure under power set also guarantees that α is a limit ordinal, in which case V_α is also closed under pairing and union, so V_α satisfies the pairing and union axioms. Finally, any V_α satisfies foundation, so V_α satisfies all the ZF axioms.

It follows that V_α also satisfies any logical consequence of the ZF axioms; that is, any proposition provable in ZF set theory. But now suppose we take the *least* α such that V_α is inaccessible. It follows that any V_β in V_α is *not* inaccessible, so V_α satisfies the sentence "there is no inaccessible V_β." Existence of an inaccessible set is therefore *not* provable in ZF.

This explains the surprising claim made at the end of Sect. 3.8: *if inaccessible sets exist, then their existence is not provable in* ZF.

It is actually in the nature of strong axiom systems like ZF that there are many sentences they can state, but neither prove nor disprove. Such sentences are said to be *independent* of the system in question. The famous incompleteness theorem of Gödel (1931), mentioned in Sect. 6.6, gives a general explanation for this phenomenon of independent sentences. ZF is particularly remarkable because its independent sentences include very natural ones for which independence can be established without appealing to the Gödel incompleteness machinery. They include the existence of "large" sets, such as inaccessibles—as was first noticed by Kuratowski (1924)—but also the axiom of choice (AC) and the continuum hypothesis (CH). The independence of AC and CH was established by a combination of the works of Gödel (1939) and Cohen (1963).

It should also be mentioned that for any "sufficiently strong" axiom system Σ there is a sentence Con(Σ), expressing the consistency of Σ, which is independent of Σ if Σ is consistent. This result is known as *Gödel's second incompleteness theorem*. It means that when we use the axioms of a strong system, such as ZF, we not only assume the axioms but also their consistency. This is natural enough, I suppose. But it means that when we claim that a sentence of ZF is independent we really should add "assuming that ZF is consistent." Because of this, statements about consistency of strong systems take a relative form: "if Σ is consistent then so is Σ'."

For example, the results of Gödel (1939) have the form:

If ZF is consistent, then so is ZF+AC+CH.

This is enough to guarantee that there is no harm in using AC or CH on top of ZF. No contradiction will result, unless there is already a contradiction in ZF.

Another relative consistency result, due to Solovay (1970), shows that inaccessibles affect the fundamental problem of measuring sets of real numbers, raised in Sect. 1.7:

If ZF+AC+"an inaccessible set exists" is consistent,
then so is ZF+"all sets of real numbers are measurable."

Surprisingly, it is not possible to prove the consistency of the latter theory from Con(ZF) alone; one really needs the extra strength derived from the assumption of an inaccessible set. Under this assumption, Solovay constructs a *model* of ZF+"all sets of real numbers are measurable"; that is, a collection of sets satisfying all the ZF axioms and in which all sets of reals are measurable. Another remarkable feature of Solovay's model (already mentioned in Sect. 5.7) is that it has the perfect set property for all sets of reals. The concept of measurability in Solovay's model is a very broad one—known as the *Lebesgue measure*—which we will study in Chap. 9.

6.9   Historical Remarks

The ordinal numbers were introduced by Cantor (1883) as a natural extension of the positive integers $1, 2, 3, \ldots$. At first Cantor was interested in the ordinals with countably many predecessors, such as

$$\omega, \omega + 1, \ldots, \omega \cdot 2, \ldots, \omega^2, \ldots,$$

which he needed to count the iterations of his derived set operation. To create these numbers he appealed to two "generating principles":

1. Forming the successor of any ordinal.
2. Forming the "limit," or least upper bound, of any set of ordinals with no greatest member.

Cantor was unclear about how a set of ordinals might be specified. But, applied to the "set of ordinals with countably many predecessors," his second generating principle produces a spectacular result: the *least uncountable ordinal*.

In this way, Cantor discovered a new path to uncountable sets. Indeed, he believed that his second generating principle would produce ordinals of higher and higher cardinality—giving a "scale" by which it might be possible to measure the cardinality of other uncountable sets, such as $\mathbb{R}$.

This was how Cantor arrived at his second, and stronger, form of the continuum hypothesis: $\mathbb{R}$ has the same cardinality as the first uncountable ordinal. He was at first optimistic that his theory of ordinal numbers would enable him to prove the continuum hypothesis, but the problem was harder than he expected, and there was a hiatus in his set theory research until the 1890s.

In the meantime, Dedekind (1888) published his theory of natural numbers in a small book *Was sind und was sollen die Zahlen?* (What are numbers and what are they for?). As mentioned in Sect. 2.2, his book was in part a rediscovery of Grassmann's inductive foundations for arithmetic, but Dedekind went further by establishing a set-theoretic foundation for induction itself. In particular, in his Theorem 126, Dedekind proved the first theorem asserting the *existence* of functions defined by induction. His proof, by piecing together partial functions, is the ancestor of many similar arguments, such as the one used in Sect. 6.3 to prove that well-ordered sets are isomorphic to ordinals.

With these results, Dedekind went further than any of his contemporaries in building set-theoretic foundations for mathematics. But in one respect he went too far—in his Theorem 66: *There exist infinite systems*. Dedekind argued that the realm S of his own thoughts is infinite, because for any thought s there is the thought

$$\varphi(s) = \text{"}s \text{ can be thought"}.$$

Since not every thought is of the form $\varphi(s)$, φ is a bijection between S and a proper subset of itself. Hence S is infinite. QED.

One problem with this argument, of course, is that S is not well-defined by mathematical standards. A deeper problem, which Dedekind did not foresee, is that *even well-defined properties may not define sets*, as mathematicians were about to learn in the 1890s.

As we saw in Sect. 3.8, in 1891 Cantor discovered that any set has more subsets than elements, so there is no largest set. Cantor was pleased with this discovery, because it put his 1883 belief in the ever-growing scale of ordinal numbers on a sound basis. But it was bad news for mathematicians who thought that every property determines a set. With no largest set, there is no "set of all sets," and hence there is no set corresponding to the property of being a set. Dedekind was disturbed by Cantor's discovery, to the extent that he withdrew a new edition of *Was sind und was sollen die Zahlen?* that was due to be published in 1903. (See Ferreirós 1999, p. 296.)

Cantor was not disturbed; in fact he tried to profit from the related result that there is no set Ω of all ordinals. He hoped to use this fact to prove the *well-ordering theorem* that every set can be well-ordered. His erroneous (and unpublished) argument is described in Ferreirós (1999), p. 295. Suppose, for the sake of contradiction, that V is a set with no well-ordering. It seemed to Cantor that V must then be so large that any ordinal, and hence Ω itself, can be mapped into V. But in that case V is a contradictory set like Ω.

Zermelo was the first to notice a flaw in the details of Cantor's argument: an unconscious use of what we now call the *axiom of choice* when mapping ordinals into V. The axiom of choice had been used several times in set theory and analysis before this, as we will see in the next chapter. Zermelo was the first to bring it to light, and in Zermelo (1904) he proved that the well-ordering theorem is actually *equivalent* to the axiom of choice. Since 1904 the axiom has played an important role in set theory, as the principle underlying many results not provable from the ZF axioms alone.

As mentioned in Sect. 6.5, most of the ZF axioms were proposed by Zermelo (1908). It is thought by some historians that Zermelo's motive was to establish foundations for his proof of the well-ordering theorem, but his declared intention was to avoid paradoxes, such as the "set of all sets." The Zermelo axioms do this in a natural way by building sets cumulatively from the bottom up: starting from the empty set and generating all other sets by pairing, union, power set, and the "definable subset" axiom, which Zermelo called *Ausseronderung* ("cutting out"). Aussonderung asserts that, for any set X and any well-defined property P, the members of X with property P form a set. Thus, properties *can* define sets, but only as subsets of sets already defined. Because of this, the existence of an infinite set has to be an axiom—not a theorem as Dedekind had hoped.

Zermelo set theory cannot prove the existence of sets of high rank. As Fraenkel (1922) observed, it cannot prove the existence of the set

$$\{\mathbb{N}, \mathcal{P}(\mathbb{N}), \mathcal{P}(\mathcal{P}(\mathbb{N})), \ldots\},$$

which is the range of the function f defined on $\mathbb{N}$ by

$$f(1) = \mathbb{N}, \quad f(n+1) = \mathcal{P}(f(n)).$$

This is one of the reasons why Fraenkel strengthened the Zermelo axioms with the replacement axiom. The schema of the replacement axiom generalizes Aussonderung, while still being in the spirit of building all sets from those previously constructed.

Fig. 6.3 Ernst Zermelo, Abraham Fraenkel, and John von Neumann

The paper of von Neumann (1923) helped to popularize the ZF system with his elegant definition of ordinals, and proofs of the basic results about them. These included the theorem that every well-ordered set is isomorphic to an ordinal and the transfinite generalization of Dedekind's theorem on definition by induction. Finally, von Neumann (1929) cemented the relationship between ZF and the cumulative set concept by using the foundation axiom to prove that every set belongs to some V_α.

The picture of John von Neumann in Fig. 6.3 is a 1925 photograph from the John von Neumann Collection, Archives of American Mathematics at SRH (item e_math_00134 from Box 4RM51). It is at the Dolph Briscoe Center for American History in Austin, Texas, and is used with their permission.

Chapter 7
The Axiom of Choice

Preview

The ZF axioms allow us to assert the existence of any set whose members are selected according to some definable "rule"—this is essentially what the replacement schema says. However, we often want to assert the existence of a set *without* knowing a rule for selecting its members. Typically, the members are simply "chosen" from other sets, but not according to any "rule." When infinitely many choices are required, we may not be able to guarantee the existence of the set without some *axiom of choice*.

The full axiom of choice (AC), described in Sect. 7.1, is a powerful axiom that greatly simplifies set theory. In particular, it implies that any set can be well-ordered, so that methods such as induction—previously applicable only to countable sets— apply to all sets.

On the other hand, AC also has some negative consequences, inasmuch as it implies the existence of sets with irregular or even bizarre properties. We give one example of an irregular property (an *undetermined* set) in Sect. 7.6, and further examples occur in Chap. 9.

For this reason it is also of interest to consider weaker axioms of choice, with consequences that are entirely positive. One of these, the *countable axiom of choice* (countable AC), is of particular value in analysis, because it is strong enough to prove some desirable properties of sets and functions, but too weak to admit the bizarre consequences of full AC. To illustrate what we mean by "positive" consequences of choice, we begin this chapter with some applications of countable AC.

7.1 Some Naive Questions About Infinity

In the early days of set theory the following questions arose and, seemingly, were easily answered.

J. Stillwell, *The Real Numbers: An Introduction to Set Theory and Analysis*, Undergraduate Texts in Mathematics, DOI 10.1007/978-3-319-01577-4_7, © Springer International Publishing Switzerland 2013

1. *Does every infinite set have a countably infinite subset?*

The naive answer is yes, because if S is infinite we can remove an element s_1 from S, and $S - \{s_1\}$ is still infinite. Then we can remove an element s_2 from $S - \{s_1\}$ and $S - \{s_1, s_2\}$ is still infinite; and so on. Proceeding in this way, we can remove an infinite sequence $s_1, s_2, s_3, \ldots$ from S, so $\{s_1, s_2, s_3, \ldots\}$ is a countably infinite subset of S.

2. *Is a countable union of countable sets countable?*

Again, the naive answer is yes. If $\{S_1, S_2, S_3, \ldots\}$ is a countable set of countable sets, let

$$S_1 = \{s_{11}, s_{12}, s_{13}, \ldots\}$$

$$S_2 = \{s_{21}, s_{22}, s_{23}, \ldots\}$$

$$S_3 = \{s_{31}, s_{32}, s_{33}, \ldots\}$$

$$\vdots$$

Then we can enumerate the members s_{ij} of the union of these sets S_k in the same way that we enumerate the members (i, j) of $\mathbb{N} \times \mathbb{N}$, namely:

$$S_1 \cup S_2 \cup S_3 \cup \cdots = \{s_{11}, s_{21}, s_{12}, s_{31}, s_{22}, s_{13}, \ldots\}.$$

Hence $S_1 \cup S_2 \cup S_3 \cup \cdots$ is countable.

3. *If a function f is sequentially continuous at x, is f continuous at x?*

We call a function f *sequentially continuous* at x if $f(x_i) \to f(x)$ for every sequence $\langle x_1, x_2, x_3, \ldots \rangle$ such that $x_i \to x$. As we know from Sect. 4.2, f is continuous at x if, for each $\varepsilon > 0$ there is a $\delta > 0$ such that

$$|x' - x| < \delta \Rightarrow |f(x') - f(x)| < \varepsilon.$$

So, given ε, we want to use sequential continuity to find a δ. Well, the alternative is that, for some ε_0, there is *no* δ. In this case we can find an x_1' with

$$|x_1' - x| < 1/2 \quad \text{and} \quad |f(x_1') - f(x)| \geq \varepsilon_0$$

then an x_2' with

$$|x_2' - x| < 1/4 \quad \text{and} \quad |f(x_2') - f(x)| \geq \varepsilon_0,$$

then an x_3' with

$$|x_3' - x| < 1/8 \quad \text{and} \quad |f(x_3') - f(x)| \geq \varepsilon_0,$$

and so on. We therefore have a sequence $\langle x_1', x_2', x_3', \ldots \rangle$ with $x_i' \to x$ but with $f(x_i') \nrightarrow f(x)$, contrary to sequential continuity.

What these examples have in common is an *infinite sequence of choices*: in the first example we infinitely often choose a new member from the set S, in the second we choose an enumeration of each set S_i, and in the third we choose infinitely many real numbers x_i'. This may seem like the proof of the Bolzano–Weierstrass theorem in Sect. 3.6, where we chose an infinite sequence of intervals $I_n \subseteq [0, 1]$, but there is an important difference. In the proof of the Bolzano–Weierstrass theorem we were able to *define* the sequence $\langle I_1, I_2, I_3, \ldots \rangle$, by taking I_n to be the leftmost half of I_{n-1} that contains infinitely many points of the given set $S \subseteq [0, 1]$.

In the three examples above the sequence of choices comes with no apparent definition—one just has to believe that it exists. Around 1900, it gradually became clear that an *axiom of choice* should be built into set theory to support all cases where a set is claimed to exist by virtue of an infinite sequence of choices. This was done by Zermelo (1904), who in fact proposed a stronger axiom allowing any infinite *set* of choices. There are many ways to state Zermelo's axiom of choice (AC), some of which we study later, but the most convenient to begin with is the following:

Axiom of Choice. *If X is any set whose members are nonempty, then there exists a function F, called a* choice function *for X, such that $F(x) \in x$ for each $x \in X$.*

Thus, the function F "chooses" a member $F(x)$ from each member x of X. Here is how the axiom of choice is deployed in the three examples above.

1. Let X be the set of all nonempty subsets of the given infinite set S, and let F be a choice function for X. Then we can *define* the sequence $\langle s_1, s_2, s_3, \ldots \rangle$ inductively in terms of F:

$$s_1 = F(S), \quad s_n = F(S - \{s_1, s_2, \ldots, s_{n-1}\}).$$

2. Let $E(S_i)$ be the set of all enumerations of the countably infinite set S_i, let

$$X = \{E(S_1), E(S_2), E(S_3), \ldots\},$$

and let F be a choice function for X. Then we can define the enumeration $\{s_{i1}, s_{i2}, s_{i3}, \ldots\}$ of S_i as $F(E(S_i))$ and complete the proof as before.

3. Our assumption is that, in any open interval I that contains x there exists an x' with $|f(x') - f(x)| \geq \varepsilon_0$. We therefore have, for each n, a nonempty set

$$J_n = \{x' : |x' - x| < 1/2^n \text{ and } |f(x') - f(x)| \geq \varepsilon_0\}.$$

We define

$$X = \{J_1, J_2, J_3, \ldots\}$$

and let F be a choice function for X. Then $x_n' = F(J_n)$ has the property

$$|x'_n - x| < 1/2^n \quad \text{and} \quad |f(x'_n) - f(x)| \geq \varepsilon_0,$$

as required for the proof.

In the latter two cases we are using the so-called *countable choice axiom*, where choices are made from each member of a countable set. In the first case we use the so-called *dependent choice axiom*, where a sequence of choices is made, each dependent on the one before. Countable choice and dependent choice are the most common choice principles used in analysis. As is clear from the examples above, it is hard to do without these principles, and they seem natural and harmless.

The full axiom of choice has some useful consequences, as we will see in the next section, but also some consequences that are *not* convenient for analysis, as we will see in the next chapter. For this reason, we will be careful to distinguish between the full axiom and weaker forms (such as dependent and countable choice) in this book. Our understanding of the real numbers turns out to depend very much upon the strength of choice principles we assume.

Exercises

Recall, from Sect. 3.9, Dedekind's definition of an infinite set: S is infinite if there is a bijection $f : S \to T$, where T is a proper subset of S. The following exercises show that this property is equivalent to the existence of a countably infinite subset of S.

7.1.1 Show that, if $s \in S - T$, then $\{s, f(s), f(f(s)), \ldots\}$ is a countably infinite subset of S.

7.1.2 Show, conversely, that a countably infinite subset of S gives a bijection $f : S \to T$, where T is a proper subset of S.

In ZF set theory it is not provable that every infinite subset of $\mathbb{R}$ contains a countable subset. We are therefore free to explore the possibility of infinite subsets of $\mathbb{R}$ without countable subsets. This turns out to throw light on the difference between sequential continuity and ordinary continuity.

7.1.3 Suppose that $S \subset \mathbb{R}$ is infinite but with no countable subset. Explain why this gives an infinite $T \subset [0, 1]$ with no countable subset.

7.1.4 Let $T \subset [0, 1]$ be the set mentioned in Exercise 7.1.3, and let t be a limit point of T, given by the Bolzano–Weierstrass theorem. Explain why we can assume $t \notin T$, without loss of generality.

7.1.5 If t and T are as in Exercise 7.1.4, show that t is not the limit of any sequence $t_1, t_2, t_3, \ldots \in T$.

7.1.6 Show the characteristic function of T is sequentially continuous at $x = t$, but not continuous there.

7.2   The Full Axiom of Choice and Well-Ordering

As we saw in the previous section, the axiom of choice allows us to make inductive constructions involving an infinite sequence of choices. So far, we have done this only for countable sequences, but there is nothing to stop us continuing through ordinal number stages until the task is complete. The most famous application of this idea is the following theorem of Zermelo (1904).

Well-ordering Theorem. *Assuming the full axiom of choice, there is a bijection between any set X and an ordinal.*

Informal proof. The basic idea could not be simpler: repeatedly choose elements $x_0, x_1, x_2, \ldots$ from X, assigning them ordinal numbers as subscripts. When all the ordinals less than α have been assigned, the next element chosen is assigned subscript α. For example, once we have chosen elements x_n for all natural numbers n, the next element chosen (if any remain) is called x_ω.

Since *any* set of ordinals has a least upper bound α, we can continue assigning ordinals to members of X until X is exhausted. This gives a bijection between X and some ordinal α (the least upper bound of the ordinals assigned to members of X).

Proof. Now we formalize the above idea with the help of a choice function F for the nonempty subsets of X. F enables us to define the following function g, mapping ordinals into X, by induction:

$$g(0) = x_0 = F(X),$$

and if $g(\beta)$ has been defined for $\beta < \alpha$, let

$$g(\alpha) = x_\alpha = F(X - \{x_\beta : \beta < \alpha\}).$$

To see that each member of X equals x_β for some ordinal β, consider the members of X that *are* of the form $g(\beta) = x_\beta$. These form a subset S of X (by the "Aussonderung" axiom), and hence

$$\{\beta : g(\beta) \text{ is defined}\} = g^{-1}(S)$$

is a set of ordinals, by the replacement schema. But this set has an upper bound α, since any set of ordinals has a least upper bound. Hence $g(\alpha) = x_\alpha$ is defined, unless the elements x_β, for $\beta < \alpha$, include all elements of X.

Since $g(\alpha)$ is *not* defined, by definition, it follows that $S = X$, and that g is a bijection between X and the ordinal α.   □

This theorem is called the *well-ordering* theorem because it says that any set can be ordered like the ordinal numbers, which are *well-ordered* by the $<$ relation as defined in Sect. 6.3:

1. The relation $<$ is a *linear* order: that is,

   for any ordinal α, $\alpha \not< \alpha$,

   for any two ordinals α and β, either $\alpha < \beta$ or $\beta < \alpha$,

   for any three ordinals α, β, and γ, if $\alpha < \beta$ and $\beta < \gamma$ then $\alpha < \gamma$.

2. Any set of ordinals has a least member in the ordering by $<$.

Well-ordering carries over to any set X if we label its elements with ordinal number subscripts as in the above proof, and then order its elements by the relation $\prec$ defined by

$$x_\alpha \prec x_\beta \Leftrightarrow \alpha < \beta.$$

The relation $\prec$ on X then inherits the well-ordering properties from the relation $<$ on ordinal numbers. We have to use the new symbol $\prec$ because the relation $\prec$ may be entirely different from the ordinary $<$ relation on X (if $<$ makes sense on X at all).

Indeed, the enormity of the well-ordering theorem first becomes apparent when we consider the case where $X = \mathbb{R}$, the set of real numbers. The ordinary $<$ relation on $\mathbb{R}$ is certainly a linear ordering, but it is definitely *not* a well-ordering, because many sets of real numbers do not have a least member under the relation $<$; for example, the set of real numbers > 0. Thus, the well-ordering $\prec$ of $\mathbb{R}$ given by the well-ordering theorem, and ultimately by the axiom of choice, must be utterly different from the ordinary ordering $<$. Indeed it turns out that, unless we assume new axioms of set theory, it is *impossible* to define a well-ordering of $\mathbb{R}$ in the language of ZF set theory. The well-ordering of $\mathbb{R}$ given by the axiom of choice "just exists"—we cannot describe it.

The elusiveness of well-ordering is symptomatic of the elusiveness of sets obtained from the axiom of choice. It cannot be less elusive, because well-ordering of every set in fact *implies* the axiom of choice. Thus, well-ordering of all sets is equivalent to the full axiom of choice.

Well-ordering implies the axiom of choice. *If every set has a well-ordering, then every set has a choice function.*

Proof. Given a set X whose members are nonempty sets, we find a choice function for X as follows. Let Y be the union of all members x of X, and take a well-ordering $\prec$ of Y. Then each x is a subset of Y, well-ordered by the relation $\prec$. So the function defined by

$$F(x) = \prec \text{-least member of } x$$

is a choice function for X. $\square$

From now on we will often abbreviate the axiom of choice by AC.

7.2.1   Cardinal Numbers

The well-ordering theorem gives a simple way to define the cardinal *number* of each set, solving the problem we raised in Sect. 6.7. Assuming AC, each set is equinumerous with an ordinal by the well-ordering theorem, so we can make the definition

$|X|$ = cardinal number of X = least ordinal equinumerous with X.

For example, $|\mathbb{N}| = \omega$ and $|\{\text{countable ordinals}\}| = \omega_1$. When talking about cardinal numbers, it is usual, following Cantor, to use the symbolism of *alephs*: $\aleph_0 = \omega$, $\aleph_1 = \omega_1$, and so on. The aleph symbol $\aleph$ is the first letter of the Hebrew alphabet.

This seemingly redundant notation is useful because there is a *cardinal arithmetic* (reflecting size) which is different from ordinal arithmetic (reflecting order). We can use the same symbols for arithmetic operations in both if we adopt the convention of using $\aleph_0, \aleph_1, \ldots$ in cardinal arithmetic and $\omega, \omega_1, \ldots$ in ordinal arithmetic. Here is an example that illustrates this usage. In ordinal arithmetic we have

$$\omega + \omega = \omega \cdot 2 \neq \omega.$$

But in cardinal arithmetic we want

$$\aleph_0 + \aleph_0 = \aleph_0,$$

to reflect the fact that the union of two disjoint countably infinite sets is countably infinite. Another example is

$$\omega \cdot \omega = \omega^2 \neq \omega \quad \text{but} \quad \aleph_0 \cdot \aleph_0 = \aleph_0,$$

the latter reflecting the fact that $\mathbb{N}^2$ is equinumerous with $\mathbb{N}$.

We are not particularly interested in cardinal arithmetic in this book, although many of the equinumerosity results in Sects. 3.1 and 3.3 can be interpreted as equations in cardinal arithmetic (see exercises). However, we occasionally take advantage of aleph notation to state results about cardinality more concisely. In particular, we use the symbol $2^{\aleph_0}$ to denote the cardinal number of $\mathbb{R}$ and of $\mathcal{P}(\mathbb{N})$. This symbol is in keeping with the result from finite mathematics that an n-element set has 2^n subsets. Using this symbol, we can express the uncountability of $\mathbb{R}$ by

$$\aleph_0 < 2^{\aleph_0}.$$

Exercises

$\mathbb{R}$ is probably the simplest example of a set for which well-ordering is not provable in ZF. Consequently, many interesting properties of $\mathbb{R}$ are provable only by assuming some form of the axiom of choice. One such property is the existence of a *Hamel basis*—a basis for $\mathbb{R}$ as vector space over $\mathbb{Q}$. In other words, a Hamel basis is a set $H \subset \mathbb{R}$ such that:

1. Each $x \in \mathbb{R}$ has the form $x = r_1 x_1 + \cdots + r_k x_k$ for some $x_1, \ldots, x_k \in H$ and $r_1, \ldots, r_k \in \mathbb{Q}$.
2. For distinct $x_i \in H$ and $r_i \in \mathbb{Q}$, $r_1 x_1 + \cdots + r_k x_k = 0$ only if $r_1 = \cdots = r_k = 0$.

7.2.1   Deduce from the above properties that each $x \in \mathbb{R}$ is *uniquely* expressible in the form $x = r_1 x_1 + \cdots + r_k x_k$ for some $x_1, \ldots, x_k \in H$. (*)

7.2.2   Assuming a well-ordering $y_0, y_1, \ldots, y_\alpha, \ldots$ of $\mathbb{R}$, define a Hamel basis of $\mathbb{R}$ by transfinite induction.

7.2.3   Given a Hamel basis $h_0, h_1, \ldots, h_\alpha, \ldots$ of $\mathbb{R}$, let

$$h(x) = \text{coefficient of } h_0 \text{ in the unique expression (*) for } x.$$

Show that $h(a + b) = h(a) + h(b)$ for each $a, b \in \mathbb{R}$, so h is an *additive function*, but h is discontinuous.

Find examples from Sects. 3.1 and 3.3 that illustrate the following equations of cardinal arithmetic.

7.2.4 $\aleph_0 + 1 = \aleph_0$.

7.2.5 $2^{\aleph_0} + \aleph_0 = 2^{\aleph_0}$.

7.2.6 $\aleph_0^{\aleph_0} = 2^{\aleph_0}$.

7.3   The Continuum Hypothesis

After Cantor discovered that $\mathbb{R}$ is uncountable, in the 1870s, he began to investigate other uncountable sets of real numbers, such as the Cantor set. All of the examples he found were actually of the same cardinality as $\mathbb{R}$, which led him to formulate the so-called *continuum hypothesis*. His first version of the continuum hypothesis, formulated in Cantor (1878), simply states that *every uncountable set of real numbers has the same cardinality as* $\mathbb{R}$.

Then in the 1880s he further developed his theory of ordinal numbers and well-ordered sets, and became convinced that *every set can be well-ordered*. In other words, he believed in the well-ordering theorem. He was not yet aware of any axiom (such as AC) that implies the well-ordering theorem; it is more likely that he simply wanted an orderly universe of sets, and this is hardly possible without the well-ordering theorem.

Pursuing this train of thought further, it would be convenient if $\mathbb{R}$ could be well-ordered, and best of all if $\mathbb{R}$ had the *smallest* uncountable cardinality, $\aleph_1$. This was Cantor's second version of the continuum hypothesis, formulated in Cantor (1883), and it is what is meant by the continuum hypothesis today. In terms of ordinal numbers, it is stated as follows:

Continuum Hypothesis. *There is a bijection between $\mathbb{R}$ and the least uncountable ordinal, ω_1.*

In the language of cardinal arithmetic: $2^{\aleph_0} = \aleph_1$.

This form of the continuum hypothesis has the advantage of making the cardinality of $\mathbb{R}$ as simple as possible, but how plausible is it? We reiterate what we said in the previous section: *unless we assume new axioms of set theory, it is impossible to define a well-ordering of $\mathbb{R}$ in the language of ZF set theory.* In the absence of plausible new axioms, AC can only guarantee the existence of a well-ordering of $\mathbb{R}$; it cannot name any such ordering.

It so happens that there is an axiom, called the *axiom of constructibility*, which provides a definable well-ordering of every set and is consistent with the ZF axioms. This remarkable new axiom was introduced by Gödel (1939) and it gives a definition of each set in a language $\mathcal{L}$ that includes the symbols of the ZF language plus symbols for all the ordinals. The sets named by $\mathcal{L}$ are called the *constructible sets*. The *axiom of constructibility* states that every set is constructible, and this axiom is consistent with ZF (roughly) because $\mathcal{L}$ gives names for enough sets to satisfy the ZF axioms. Moreover, it is possible to define a well-ordering of all the formulas in $\mathcal{L}$, and hence of all the constructible sets. It follows that each constructible set gets a well-ordering, since all of its members are constructible sets. This means that the universe L of constructible sets satisfies not only the ZF axioms but also the well-ordering theorem, and hence the axiom of choice.

Moreover, it turns out (by no means obviously), that each real number in L is defined by a formula in $\mathcal{L}$ involving only symbols for *countable* ordinals. It follows from this that the real numbers in L can be ordered in a sequence of length ω_1, and hence that L satisfies the continuum hypothesis. Thus, L is a model of the ZF axioms, plus the axiom of choice and the continuum hypothesis. It follows that the latter two propositions are *consistent with* the axioms of ZF. This is how Gödel (1939) showed that the axiom of choice and the continuum hypothesis *could not be disproved* from the ZF axioms, though it remained to be seen whether they could be proved.[1]

In any case, while a definable well-ordering may be good for the universe, it is not necessarily good for $\mathbb{R}$. The definable well-ordering of $\mathbb{R}$ implied by the axiom of constructibility implies that there are definable subsets of $\mathbb{R}$ with bizarre properties, as we will see in this chapter and in Chap. 9. Also, the axiom of constructibility limits the size of sets that can exist, and modern set theory often requires sets larger than the axiom of constructibility will allow. For these reasons, mathematicians have not rushed to add the axiom of constructibility to the ZF axioms. It remains a delicate matter to decide how ZF should be strengthened to provide the clearest possible picture of $\mathbb{R}$. In the remainder of this book we will study how our view of $\mathbb{R}$ depends upon which axioms are adopted.

Exercises

The following exercises explore part of Gödel's proof that the continuum hypothesis is consistent with ZF: the cardinality of sets of names in the language $\mathcal{L}$.

7.4.1 Given an infinite countable ordinal γ, and assuming that there are countably many symbols in the language of ZF, explain how to use the ordinals $< \gamma$ to encode the symbols of ZF *plus* symbols for all the ordinals $< \gamma$.

---

[1] As mentioned in Sect. 6.8, Cohen (1963) showed that the axiom of choice and the continuum hypothesis cannot be proved from the ZF axioms. It follows that the axiom of constructibility is not provable in ZF either.

7.4.2 Assuming the encoding of symbols from Exercise 7.4.1, formulas in the language of ZF
plus symbols for the ordinals $< \gamma$ become finite sequences $\langle \alpha_1, \ldots, \alpha_n \rangle$ of ordinals less than
γ. Describe a well-ordering of such sequences and explain why its order type is countable.

7.4.3 Now consider formulas in the language of ZF plus symbols for all countable ordinals. (As
mentioned above, Gödel proved that each constructible real number can be defined by such
a formula.) Deduce from Exercise 7.4.2 that the set of all such formulas can be well-ordered,
with order type ω_1.

7.4 Filters and Ultrafilters

No axiom of choice is needed to prove results about natural numbers, since the
induction axiom ensures that $\mathbb{N}$ is well-ordered. The lowest level theorems for which
an axiom of choice may be required are those about *sets* of natural numbers. In this
section we will give an example—the *existence of a nonprincipal ultrafilter over*
$\mathbb{N}$—which also turns out to be interesting in analysis (see Chap. 9). The ultrafilter
example is also interesting because it involves an uncountable infinity of choices.
Such uses of AC can lead to strange results, and we will see in Chap. 9 that this is
one of them.

We will take filters and ultrafilters to be certain collections of subsets of $\mathbb{N}$,
though the definition applies to subsets of any set.

Definition. A collection $\mathcal{F}$ of subsets is called a *filter* if

1. $\emptyset \notin \mathcal{F}$.
2. If $A \in \mathcal{F}$ and $A \subseteq B$ then $B \in \mathcal{F}$.
3. If $A, B \in \mathcal{F}$ then $A \cap B \in \mathcal{F}$.

In other words, a filter is a collection of subsets that does not include the empty set
and is closed under supersets and intersections.

The reason for the name "filter" is that if a set A is "caught" in $\mathcal{F}$, then so is any
set B larger than A. Two important examples of filters on $\mathbb{N}$ are the following:

- For any $a \in \mathbb{N}$, $\mathcal{F}_a = \{A \subseteq \mathbb{N} : a \in A\}$ is a filter. $\mathcal{F}_a$ is called the *principal* filter
  generated by a.
- The set of *cofinite* subsets of $\mathbb{N}$, $\{X \subseteq \mathbb{N} : \mathbb{N} - X \text{ is finite}\}$, is a filter.

In both of these examples it is easy to check that conditions 1, 2, 3 for a filter are
satisfied. The principal filter satisfies an additional condition that makes it what we
call an *ultrafilter*:

4. For each $B \subseteq \mathbb{N}$, either $B \in \mathcal{F}$ or $\mathbb{N} - B \in \mathcal{F}$.

This raises the question: are there any *non*principal ultrafilters over $\mathbb{N}$? One way
to answer this question would be to extend the filter of cofinite sets to an ultrafilter:
there is obviously no $a \in \mathbb{N}$ that belongs to all cofinite sets, so any ultrafilter
containing the cofinite sets is not principal. In fact, we will show that any filter
can be extended to an ultrafilter, by making uncountably many applications of the
following result.

Filter Extension. *If $\mathcal{F}$ is a filter over $\mathbb{N}$ that is not an ultrafilter—so both $A \notin \mathcal{F}$ and $\mathbb{N} - A \notin \mathcal{F}$ for some A—then there is a filter $\mathcal{H} \supset \mathcal{F}$ with $A \in \mathcal{H}$.*

Proof. We extend the set $\mathbb{F}$ to a set $\mathcal{H}$ in two stages that ensure closure of $\mathcal{H}$ under intersections and supersets.

Stage 1. Add to $\mathcal{F}$ all sets of the form $A \cap F$, where $F \in \mathcal{F}$. The resulting set

$$\mathcal{G} = \mathcal{F} \cup \{A \cap F : F \in \mathcal{F}\}$$

is closed under intersections, as one sees by forming the intersections of different kinds of members: if $F_1, F_2 \in \mathcal{F}$ then $F_1 \cap F_2 = F \in \mathcal{F}$ by closure of $\mathcal{F}$, $(A \cap F_1) \cap F_2 = A \cap (F_1 \cap F_2) = A \cap F \in \mathcal{G}$, and $(A \cap F_1) \cap (A \cap F_2) = A \cap (F_1 \cap F_2) \in \mathcal{G}$ likewise.

Stage 2. Add all supersets $B \supseteq A \cap F$ of the sets added at Stage 1. Since the supersets of each $F \in \mathcal{F}$ were already in $\mathcal{F}$, the resulting set

$$\mathcal{H} = \mathcal{F} \cup \{B \supseteq A \cap F : F \in \mathcal{F}\}$$

is closed under supersets. It is also closed under intersections, as we again see by cases. For example, if $B_1 \supseteq A \cap F_1$ and $B_2 \supseteq F_2$, then $B_1 \cap B_2 \supseteq A \cap (F_1 \cap F_2)$, so $B_1 \cap B_2$ is one of the supersets of an $A \cap F$, already included.

Finally, we observe that the empty set $\emptyset \notin \mathcal{H}$ because the elements of $\mathcal{H}$ are supersets of sets of the form F or $A \cap F$, where $F \in \mathcal{F}$. We know that each $F \neq \emptyset$, because $\mathcal{F}$ is a filter. And if $A \cap F = \emptyset$ then $F \subseteq \mathbb{N} - A$, which implies $\mathbb{N} - A \in \mathcal{F}$ (by closure under supersets), contrary to assumption. $\square$

Filter extension leads us to believe that any filter $\mathcal{F}$ that is not an ultrafilter can be extended to an ultrafilter $\mathcal{U}$ by finding a set A such that $A \notin \mathcal{F}$ and $\mathbb{N} - A \notin \mathcal{F}$ and extending $\mathcal{F}$ to include it, then iterating this process "until no such sets remain." This is the kind of infinite process that AC enables us to carry out. Each single step of extending $\mathcal{F}$ to include A will be called *extension of $\mathcal{F}$ by A*.

Extension to an Ultrafilter. *Any filter over $\mathbb{N}$ is contained in an ultrafilter over $\mathbb{N}$.*

Proof. Given a filter $\mathcal{F}$ over $\mathbb{N}$ we build an increasing sequence of filters $\mathcal{F}_\alpha$ whose union is an ultrafilter. To define the $\mathcal{F}_\alpha$ we use AC to obtain a well-ordering $A_0, A_1, \ldots, A_\alpha, \ldots$, for $\alpha < \lambda$, of all the subsets of $\mathbb{N}$. Then we let

$$\mathcal{F}_0 = \mathcal{F}$$

$$\mathcal{F}_{\alpha+1} = \begin{cases} \mathcal{F}_\alpha \text{ if } A_\alpha \in \mathcal{F}_\alpha \text{ or } \mathbb{N} - A_\alpha \in \mathcal{F}_\alpha \\ \text{extension of } \mathcal{F}_\alpha \text{ by } A_\alpha \text{ otherwise} \end{cases}$$

$$\mathcal{F}_\beta = \bigcup_{\gamma < \beta} F_\gamma \text{ for each limit ordinal } \beta.$$

It follows by filter extension that $\mathcal{F}_{\alpha+1}$ is a filter when $\mathcal{F}_\alpha$ is. It is also clear that $\bigcup_{\gamma < \beta} F_\gamma$ is a filter when each $\mathcal{F}_\gamma$ is: $\bigcup_{\gamma < \beta} F_\gamma$ is closed under intersection and supersets because each $\mathcal{F}_\gamma$ is, and $\emptyset \notin \bigcup_{\gamma < \beta} F_\gamma$ because $\emptyset \notin$ each $\mathcal{F}_\gamma$.

Thus, it follows by transfinite induction that $\mathcal{F}_\alpha$ is a filter for each α, and $\mathcal{F}_\lambda$ is also a filter by the argument for limit ordinals. Finally, either $A_\alpha \in \mathcal{F}_\lambda$ or $\mathbb{N} - A_\alpha \in \mathcal{F}_\lambda$ for each α, by construction, so $\mathcal{F}_\lambda$ is an ultrafilter that extends $\mathcal{F}$.   □

In Sect. 7.8 below we will give another proof that each filter extends to an ultrafilter, again using filter extension, but replacing the transfinite induction by another form of AC.

Exercises

7.4.1 Explain why $\mathcal{F} = \{A \subseteq \mathbb{N} : 0 \in A \text{ and } 1 \in A\}$ is a filter but not an ultrafilter.

7.4.2 Show that the filter of cofinite subsets of $\mathbb{N}$ is countable.

7.4.3 Show that the complements $\mathbb{N} - F$ of the sets F in a filter $\mathcal{F}$ form a *Boolean algebra ideal*; that is, a set $\mathcal{I}$ that is closed under unions and under intersections with arbitrary sets $\subseteq \mathbb{N}$. If $\mathcal{F}$ is an ultrafilter, show that $\mathcal{I}$ is a maximal ideal.

7.4.4 Interpret a nonprincipal ultrafilter $\mathcal{F}$ over $\mathbb{N}$ as a 0–1 measure μ on $\mathcal{P}(\mathbb{N})$. That is, if we set $\mu(a) = 1$ for $A \in \mathcal{F}$ and $\mu(A) = 0$ for $A \notin \mathcal{F}$, show that we have a measure on all sets $A \subseteq \mathbb{N}$ which is *additive* in the sense that $\mu(A_1 \cup A_2) = \mu(A_1) + \mu(A_2)$ for disjoint A_1, A_2.

7.5 Games and Winning Strategies

Our next application of AC is in the theory of infinite games. To put this result in context we should first say something about *finite games with perfect information*. We consider two-person games, in which players I and II move alternately and there is a bound on the length of complete sequences of moves ("plays" of the game). Such a game is called finite, and it is said to be *with perfect information* if each player knows all previous moves. Typical games without perfect information are card games, where a player does not initially know the cards another player has been dealt, and typical games with perfect information are tic-tac-toe and chess.

In a two-person game with perfect information one of the players may have a *winning strategy*; that is, a rule for making moves that always leads to a win. In tic-tac-toe, neither player has a winning strategy, because the game can end in a draw. But if we change the rules so that (for example) a draw counts as a win for player I, then one of the players does have a winning strategy. This is just one instance of a remarkably general, yet simple, theorem:

Winning strategy theorem. *If $\mathcal{G}$ is a finite two-person game with perfect information, in which every play ends in a win for one of the players, then either player I or player II has a winning strategy.*

Proof. Since the game $\mathcal{G}$ is finite, all possible plays of $\mathcal{G}$ can be captured as paths in a tree like that shown in Fig. 7.1.

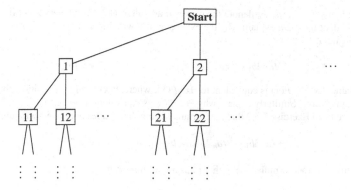

Fig. 7.1 The tree of plays of a game

The vertex **Start** represents the starting position of the game, the vertices 1, 2, ... below it represent the positions that can reached by the first move (which is by I), the vertices 11, 12, ... and 21, 22, ... below these represent the positions that can be reached by the second move (which is by II), and so on. Since G is finite, there is a maximum value N for the length of downward branches from the **Start** vertex.

We now prove the existence of a winning strategy for all such games G, by induction on N. If $N = 1$ then the game ends in one move and, by the hypothesis of the theorem, every move leads to a win for either I or II. If any move leads to a win for I, then choosing that move is a winning strategy for I. If every move leads to a win for II, then letting I make any move is a winning strategy for II. This completes the base step of the induction.

Now, for the induction step, suppose that either I or II has a winning strategy for any game of length $< N$. Among such games are the *sub*games of G whose starting positions are the vertices 1, 2, ... in the tree of plays of G. Thus, each of the latter games has a winning strategy for either I or II.

But then I or II has a winning strategy for G itself. If any of the games with starting position 1, 2, ... has a winning strategy for I, then I has winning strategy for G. It consists of making his first move into such a subgame, n say, and thereafter playing a winning strategy for the game n. If none of the games 1, 2, ... has a winning strategy for I, then they all have winning strategies for II, in which case II has a winning strategy for G. Namely, II plays a winning strategy for whichever game n that player I moves into.

This completes the induction, and the proof of the theorem.   □

Exercises

A very short proof of the winning strategy theorem may be written down using *quantifiers*: $\forall x$, meaning "for all x," and $\exists x$, meaning "there exists an x."

7.5.1 If $a_1, b_1, a_2, b_2, \ldots, a_n, b_n$ denote the moves made alternately by players I and II in a game of length at most $2n$, explain why the existence of a winning strategy for II is expressed by the formula.

$$\forall a_1 \exists b_1 \cdots \forall a_n \exists b_n (a_1, b_1, \ldots, a_n, b_n \text{ is a win for II})$$

7.5.2 Explain why $\neg \forall x P(x)$ is equivalent to $\exists x \neg P(x)$, where $P(x)$ is any proposition about x and $\neg$ means "'not." Similarly explain why $\neg \exists x P(x)$ is equivalent to $\forall x \neg P(x)$.

7.5.3 Deduce from Exercise 7.5.2 that the formula saying that II does not have a winning strategy,

$$\neg \forall a_1 \exists b_1 \cdots \forall a_n \exists b_n (a_1, b_1, \ldots, a_n, b_n \text{ is a win for II}),$$

is equivalent to a formula saying that I has a winning strategy.

7.6 Infinite Games

If we remove the restriction that all plays in a game have length bounded by some integer N, and if we allow each move to be chosen from some countable set, then the tree in Fig. 7.1 now represents all possible plays in a *countably infinite game with perfect information*. Each play in such a game is represented by an infinite sequence $\langle a_1, b_1, a_2, b_2, \ldots \rangle$, where $a_1, a_2, a_3, \ldots$ represent the successive moves by player I and $b_1, b_2, b_3, \ldots$ represent the successive moves by player II. The game itself is defined by a set X of such sequences; namely, those that represent a win for player I. We call this game $\mathcal{G}_X$. Thus, in $\mathcal{G}_X$ player I tries to ensure that the sequence $\langle a_1, b_1, a_2, b_2, \ldots \rangle \in X$, while II tries to ensure that $\langle a_1, b_1, a_2, b_2, \ldots \rangle \notin X$.

Such games were first considered by Hugo Steinhaus in 1925 and he conjectured that, by analogy with finite games, for any set X one player has a winning strategy for $\mathcal{G}_X$. A short while later, Banach and Mazur showed that the Steinhaus conjecture is false if we assume the axiom of choice. AC makes it possible to define a set X for which neither player has a winning strategy for the game $\mathcal{G}_X$. Such a set X is called *undetermined*.

The Banach–Mazur proof actually uses a set $X \subseteq [0, 1]$, and players I and II choose successive digits of a real number in $[0,1]$. It is similar, but slightly more convenient, to use $\mathcal{N} = \mathbb{N}^{\mathbb{N}}$ in place of $[0,1]$, as we do here. In fact, any ultrafilter $U \subset \mathcal{P}(\mathbb{N})$ containing the cofinite filter gives a natural example of an undetermined set $X \subset \mathcal{N}$, as we will see in the exercises.

7.6.1 Strategies

Before showing how AC gives an undetermined set, we need to define what a *strategy* is. In any game $\mathcal{G}_X$, a *play* is a sequence $\langle a_1, b_1, a_2, b_2, a_3, b_3, \ldots \rangle$, where $\langle a_1, a_2, a_3, \ldots \rangle$ is the sequence played by I and $\langle b_1, b_2, b_3, \ldots \rangle$ is the sequence played by II. A *strategy* σ is a positive integer-valued function defined on all finite sequences of positive integers, including the empty sequence $\langle \rangle$. Player I *plays* strategy σ by making the moves

$$a_1 = \sigma(\langle \rangle),$$

$$a_2 = \sigma(\langle b_1 \rangle),$$

$$a_3 = \sigma(\langle b_1, b_2 \rangle),$$

$$\vdots$$

in response to the moves $b_1, b_2, \ldots$ made by player II. Player II plays strategy σ by making the moves

$$b_1 = \sigma(\langle a_1 \rangle),$$

$$b_2 = \sigma(\langle a_1, a_2 \rangle),$$

$$b_3 = \sigma(\langle a_1, a_2, a_3 \rangle),$$

$$\vdots$$

in response to the moves $a_1, a_2, a_3, \ldots$ made by player I.

We let $\sigma * b$ denote the sequence $\langle a_1, b_1, a_2, b_2, \ldots \rangle$ that results when I plays strategy σ on the sequence b of moves made by player II. And we say that σ is a *winning strategy for* I in the game $\mathcal{G}_A$ if $\sigma * b \in A$ for all $b \in \mathcal{N}$. Similarly, we let $a * \sigma$ denote the sequence that results when II plays strategy σ on the sequence a of moves made by player I. And we say that σ is a *winning strategy for* II in game $\mathcal{G}_A$ if $a * \sigma \notin A$ for all $a \in \mathcal{N}$.[2]

It is an easy exercise to show that the set of strategies σ has the same cardinality as the set $\mathcal{N}$, namely $2^{\aleph_0}$. We use this fact in the proof below to define a set $X \subset \mathcal{N}$ in ordinal-numbered stages $\alpha < 2^{\aleph_0}$, alongside an enumeration of strategies σ_α for $\alpha < 2^{\aleph_0}$. This, of course, assumes AC to obtain a well-ordering of $2^{\aleph_0}$. Also, by taking *smallest* ordinal equinumerous with $2^{\aleph_0}$, we can assume that the set of σ_β for $\beta < \alpha$ has cardinality less than $2^{\aleph_0}$. It will also be convenient to use AC to make choices at each of the infinitely many stages α.

An undetermined set. *There exists a set $X \subset \mathcal{N}$ for which neither player has a winning strategy for the game $\mathcal{G}_X$.*

Proof. Let $\{\sigma_\alpha : \alpha < 2^{\aleph_0}\}$ be an enumeration of all strategies. Using this enumeration, we will inductively choose the members of disjoint subsets of $\mathcal{N}$,

---

[2]A reason for writing σ on different sides in the two notations is that in $\sigma * b$ we use σ *before* seeing b, namely, on the empty sequence; and in $a * \sigma$ we use σ *after* seeing (the first member of) a.

$$X = \{x_\alpha : \alpha < 2^{\aleph_0}\} \quad \text{and} \quad Y = \{y_\alpha : \alpha < 2^{\aleph_0}\},$$

as explained below. Each x_α will *witness* the fact that σ_α is not a winning strategy for II in the game $\mathcal{G}_X$, because we will arrange that $a * \sigma_\alpha = x_\alpha \in X$ for some $a \in \mathcal{N}$. Each y_α will witness the fact that σ_α is not a winning strategy for I either, because we will arrange that $\sigma_\alpha * b = y_\alpha \notin X$ for some $b \in \mathcal{N}$.

At stage 0 we make these arrangements, and keep x_0 and y_0 in disjoint sets, by letting

$$x_0 = \text{any value of } a * \sigma_0,$$

$$y_0 = \text{any value of } \sigma_0 * b \text{ unequal to } x_0.$$

Such a value y_0 exists because $\sigma_0 * b$ takes $2^{\aleph_0}$ values as b runs through $\mathcal{N}$, since b consists of all the even-numbered places in $\sigma_0 * b$. (Indeed, both $\sigma * b$ and $a * \sigma$ take $2^{\aleph_0}$ values for any fixed σ, a fact we will rely on at stage α.)

At stage α less than $2^{\aleph_0}$ values x_β, y_β have yet been chosen, so enough values $a * \sigma_\alpha$ and $\sigma_\alpha * b$ remain to let

$$x_\alpha = \text{any value of } a * \sigma_\alpha \text{ not in } \{y_\beta : \beta < \alpha\},$$

$$y_\alpha = \text{any value of } \sigma_\alpha * b \text{ not in } \{x_\beta : \beta \leq \alpha\}.$$

It follows by induction on α that X and Y have no common member, and for each $\alpha < 2^{\aleph_0}$ we have witnesses to the fact that σ_α is not a winning strategy for either II or I in the game $\mathcal{G}_X$. Since the σ_α exhaust *all* strategies, it follows that the set X is undetermined.   □

Exercises

7.6.1 Using the Cantor–Schröder–Bernstein theorem, or otherwise, show that the set of strategies has the same cardinality as $\mathcal{N}$.

7.6.2 Supposing we take $A \subseteq [0, 1]$ and let I and II alternately choose decimal digits of a number in $[0,1]$. Show that II has a winning strategy for the game with $A = \mathbb{Q}$.

7.6.3 Find a similar example $A \subset \mathcal{N}$.

7.6.4 By imitating the above construction of an undetermined set above, or otherwise, show that there is an undetermined set $X \subset [0, 1]$ for the game where I and II alternately choose decimal digits of a number.

The next group of exercises show that an undetermined set X is also obtainable from the theorem on ultrafilters in Sect. 7.4. Specifically, we use an ultrafilter U that extends the *cofinite filter*, the set of cofinite subsets of $\mathbb{N}$. The set X is defined to be the set of sequences

$$\langle x_1, x_2, x_3, x_4, \ldots \rangle \in \mathcal{N}$$

such that

$$x_1 < x_2 < x_3 < x_4 < \cdots \quad \text{and} \quad [1, x_1) \cup [x_2, x_3) \cup [x_4, x_5) \cup \cdots \in U.$$

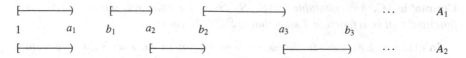

Fig. 7.2 Two members of the ultrafilter U

We suppose, for the sake of contradiction, that player I has a winning strategy σ for $\mathcal{G}_X$. That is, whatever increasing sequence $\langle b_1, b_2, b_3, \ldots \rangle$ is played by II, the sequence

$$\langle a_1, b_1, a_2, b_2, a_3, b_3, \ldots \rangle \in X,$$

when I plays strategy σ.

7.6.5   Deduce that the set $A_1 = [1, a_1) \cup [b_1, a_2) \cup [b_2, a_3) \cup \cdots \in U$ when I plays strategy σ.

Now consider the following sequence of numbers, also chosen with the help of the function σ:

$$b_2 = \sigma(\langle a_2 \rangle),$$

$$b_3 = \sigma(\langle a_2, a_3 \rangle),$$

$$b_4 = \sigma(\langle a_2, a_3, a_4 \rangle),$$

$$\vdots$$

7.6.6   Show that the play

$$\langle a_1, a_2, b_2, a_3, b_3, a_4, b_4, \ldots \rangle \quad \text{is also a win for I,}$$

and hence that the set $A_2 = [1, a_1) \cup [a_2, b_2) \cup [a_3, b_3) \cup \cdots \in U$.

Thus, we have engineered sets $A_1, A_2 \in U$ that look like Fig. 7.2.

7.6.7   Use the fact that U is an ultrafilter to deduce that

$$A_1 \cap A_2 = [1, a_1) \in U,$$

and hence that the complement of $A_1 \cap A_2$ is a cofinite set *not* in U.

This contradicts the assumption that U is an extension of the cofinite filter, so I does not have a winning strategy. We find a similar contradiction if player II has a winning strategy, hence neither player has a winning strategy for the game $\mathcal{G}_X$.

7.7   The Countable Axiom of Choice

The three questions raised in Sect. 7.1 can all be answered by the following special case of the axiom of choice—the *countable* axiom of choice. We call it countable AC for short, since we denote the full axiom of choice by AC.

Countable AC. *Any countable set* $\{S_1, S_2, S_3, \ldots\}$ *of nonempty sets* S_n *has a choice function; that is, a function* f *such that* $f(S_n) \in S_n$ *for each* n.

In question 2 we have to choose an *enumeration* of each set S_n, so we want a choice function for the set $\{S_1, S_2, S_3, \ldots\}$, where

$$S_n = \{\text{enumerations of } S_n\}.$$

In question 3 we choose a real number x'_n with $|x'_n - x| < 1/2^n$ and $|f(x'_n) - f(x)| \geq \varepsilon_0$, so we want a choice function for the set $\{S_1, S_2, S_3, \ldots\}$ where

$$S_n = \{x' : |x' - x| < 1/2^n \text{ and } |f(x'_n) - f(x)| \geq \varepsilon_0\}.$$

For question 1 it is not so clear what to do, because in the "obvious" solution each choice depends on the previous one. However, we can prescribe a suitable countable set $\{S_1, S_2, S_3, \ldots\}$ in advance by defining

$$S_n = \{n\text{-element subsets of } S\}.$$

Since S is infinite, each S_n is nonempty, so by countable AC we choose a set $f(S_n) = S_n$ from each S_n. Then the union of sets S_n is infinite, and countable by question 2.

Thus, countable AC is useful (and in fact necessary) to prove some basic theorems of analysis. However, the *full* axiom of choice, AC, is not necessary to prove the above theorems, and in fact AC causes some irregularities in the theory of $\mathbb{R}$, as we will see in the next chapter. Therefore, it is of interest to explore other axioms, strong enough to imply countable AC for subsets of $\mathbb{R}$, but less disruptive to the theory of $\mathbb{R}$ than AC.

An interesting candidate for such an axiom is the following *axiom of determinacy*, AD. We state AD for subsets of $\mathcal{N}$, or the set of irrationals in $[0,1]$, but the corresponding statement for $[0,1]$ or for $\mathbb{R}$ is equivalent.

Axiom of determinacy. *For any set* $X \subset \mathcal{N}$, *either player I or player II has a winning strategy for the game* $\mathcal{G}_X$.

AD implies countable AC for subsets of $\mathcal{N}$. *Given a countable set* $S = \{S_1, S_2, S_3, \ldots\}$, *where each* $S_n \subset \mathcal{N}$, *AD gives a choice function for* S.

Proof. Given a countable set $\{S_1, S_2, S_3, \ldots\}$ of sets $S_n \subset \mathcal{N}$, consider the following game. If

$$\text{I plays } \langle a_1, a_2, a_3, \ldots, \rangle \in \mathcal{N}$$

$$\text{and II plays } \langle b_1, b_2, b_3, \ldots, \rangle \in \mathcal{N}$$

then II wins if and only if $\langle b_1, b_2, b_3, \ldots \rangle \in S_{a_1}$. This game can be formulated as $\mathcal{G}_X$ for a certain $X \subset \mathcal{N}$, namely

$$X = \{\langle a_1, b_1, a_2, b_2, \ldots \rangle : \langle b_1, b_2, b_3, \ldots \rangle \notin S_{a_1}\}.$$

Therefore, AD says that either I or II has a winning strategy.

Now player I does *not* have a winning strategy for this game, because after I plays a_1 player II can always win by playing some $\langle b_1, b_2, b_3, \ldots \rangle$ in the nonempty set S_{a_1}. So player II has a winning strategy; that is, a function f which (among other things) for each a_1 gives a $\langle b_1, b_2, b_3, \ldots \rangle \in S_{a_1}$. In other words, a winning strategy for II gives a choice function for the sets $S_1, S_2, S_3, \ldots$. $\quad\square$

This surprising theorem tells us that, although AD is incompatible with full AC (by the previous section), it actually implies enough choice for some important applications to analysis.

Exercises

The countable AC is not provable in ZF. In fact ZF cannot prove its consequence that a countable union of countable sets is countable, or even the extreme special case that $\mathbb{R}$ is not a countable union of countable sets. Amazingly, it is consistent with ZF for $\mathbb{R}$ to be a countable union of countable sets. The need for at least countable AC in analysis is underlined by the bizarre consequences of assuming that $\mathbb{R}$ is a countable union of countable sets, which include:

7.7.1 There are countably many sets of measure 0 whose union has measure 1.
7.7.2 Every set $S \subseteq \mathbb{R}$ is a countable union of countable sets.
7.7.3 Every real function is a limit of limits of continuous functions. (*Hint*: First prove that any function with countably many nonzero values is a limit of continuous functions.)

7.8 Zorn's Lemma

We constructed an ultrafilter by transfinite induction in Sect. 7.4 because induction and ordinals are a major theme in this book, and the ultrafilter construction is a natural application of them. However, it should be pointed out that many books construct ultrafilters by a different method, called *Zorn's lemma*, which is useful for constructing many types of "maximal" objects. Briefly put, Zorn's lemma is an axiom of choice for people who dislike ordinals.

Zorn's Lemma. *Suppose that $\mathcal{T}$ is a set such that each linearly ordered subset S (under the relation of set inclusion)[3] has an upper bound: that is, an $X \in \mathcal{T}$ such that $Y \subseteq X$ for each $Y \subseteq S$. Then $\mathcal{T}$ has a maximal element: that is, a $Z \in \mathcal{T}$ such that Z is not properly contained in any other member of $\mathcal{T}$.*

---

[3]The usual statement of Zorn's lemma does not restrict the ordering to be set inclusion. However, this is the only case we need, and there is really no loss of generality.

Before proving that Zorn's lemma is equivalent to AC (or to the well-ordering theorem), we illustrate the use of Zorn's lemma by a new proof that every filter extends to an ultrafilter.

Extension to an Ultrafilter. *Any filter over* $\mathbb{N}$ *is contained in an ultrafilter over* $\mathbb{N}$.

Proof. Let $\mathcal{T}$ be the set of all filters over $\mathbb{N}$. If S is a set of filters that is linearly ordered by set inclusion, then the union X of all filters in S is itself a filter: X is closed under intersection and superset because any members of X belong to some filter in S (this is where the linear ordering is important—we cannot have a pair of members of X that do *not* belong to a single filter in S), and $\emptyset$ is not in X because $\emptyset$ is not in any member of S.

Thus, $\mathcal{T}$ satisfies the condition of Zorn's lemma, and hence $\mathcal{T}$ has a maximal element Z. In other words, Z is a filter over $\mathbb{N}$ that is not properly contained in any other filter over $\mathbb{N}$. This implies that Z is an ultrafilter, otherwise it could be extended by the filter extension theorem.   □

The main difference between this proof and the one given in the previous section is replacement of a transfinite repetition of the extension process by the single step of selecting a maximal element. This is typical of the way Zorn's lemma works: it is able to hide a transfinite extension process and, not surprisingly, this is because a transfinite extension process is built into the proof of Zorn's lemma itself.

Equivalence Theorem. *In ZF, Zorn's lemma is equivalent to* AC.

Proof. We first use AC to prove Zorn's lemma. Suppose we are given a set $\mathcal{T}$ in which each linearly ordered subset S has an upper bound. AC gives a function f such that

$$f(S) = \text{an upper bound of } S$$

for each $S \subseteq \mathcal{T}$ that is linearly ordered by $\subseteq$. Moreover, we can stipulate that

$$f(S) \supsetneq \text{each element of } S$$

if such an element exists. Using the function f, and transfinite induction, we define a linearly ordered sequence of sets $A_\alpha \in \mathcal{T}$ whose upper bound is necessarily maximal, namely, let

$$A_0 = \text{any element of } \mathcal{T},$$

$$A_\alpha = f(\mathcal{T} - \{f(\beta) : \beta < \alpha\}).$$

The A_α form a set, by the replacement axiom, since α cannot exceed the cardinality of $\mathcal{T}$. The set of A_α is linearly ordered, by transfinite induction. And its upper bound is maximal by definition of f, since f always chooses an element greater than any chosen earlier, if it can.

Conversely, if Zorn's lemma holds, we can obtain a well-ordering of any set X (and hence AC) as follows. Consider the set $\mathcal{T}$ of all bijections between subsets of X and ordinals. Such a bijection,

$$g : Y \to \alpha,$$

is of course a set (of ordered pairs), and if $g_1 \subset g_2$ then g_2 extends g_1, from a subset $Y_1 \subseteq X$ to a larger subset $Y_2 \subseteq X$, by agreeing with g_1 on Y_1 and mapping the members of $Y_2 - Y_1$ to larger ordinals.

So if we have a set S of these bijections, linearly ordered by inclusion, its union will itself be such a bijection, and hence an upper bound of S in $\mathcal{T}$. Thus, $\mathcal{T}$ satisfies the conditions of Zorn's lemma.

Zorn's lemma then gives a maximal element of $\mathcal{T}$; that is, a bijection g between a subset $Y \subseteq X$ and an ordinal α that *cannot be extended*. It follows that $Y = X$ (because if $x \in X - Y$ we can extend $g : Y \to \alpha$ by the ordered pair $\langle x, \alpha \rangle$) and hence g gives a well-ordering of X.   □

Exercises

Zorn's lemma is often used to prove the existence of maximal objects in algebra. In the following exercises we assume that the reader is familiar with the concepts of vector space, ring, ideal, and algebraically closed field.

7.8.1 Give another proof of the existence of a Hamel basis for $\mathbb{R}$ (Exercise 7.2.2) using Zorn's lemma.

7.8.2 More generally, use Zorn's lemma to prove that each vector space has a basis.

7.8.3 Use Zorn's lemma to prove that each ideal in a ring has an extension to a maximal ideal.

7.8.4 Use Zorn's lemma to prove that each field has an algebraic closure.

7.9   Historical Remarks

AC was used unconsciously for about 30 years before its explicit statement by Zermelo (1904). Many such instances are described in the book Moore (1982). The first was the proof by Cantor in 1871 that sequential continuity at a point implies ordinary continuity, a result we saw in Sect. 7.1 to depend on countable AC. This theorem is attributed to Cantor by Heine (1872), p. 182. The second unconscious use of AC was an algebraic theorem of Dedekind (1877) about modules.

In § 1 of his paper, Dedekind defines a *module* M to be a set of numbers (generally complex numbers) that is closed under addition and subtraction. He calls M a module because it leads to a notion of *congruence modulo M*; namely, for any numbers a and b,

$$a \equiv b \pmod{M} \Leftrightarrow a - b \in M.$$

It follows easily from this definition that the numbers are partitioned into disjoint *congruence classes*, where a and b belong to the same class if and only if $a \equiv b$ (mod M). In §2 of his paper, Dedekind claimed that there is a set S that includes exactly one member of each equivalence class. This seemingly obvious result, which is routine in algebra classes today, depends heavily on AC.

To appreciate why, consider $\mathbb{Q}$ as a module in $\mathbb{R}$. Then a and b belong to the same equivalence class if and only if $a - b$ is rational, so the congruence classes are very easy to understand. But a set S with exactly one member from each equivalence class is obtainable only with the help of AC—countable AC does not suffice—and no explicit definition of S can be given. S even has the bizarre property of being *nonmeasurable*, as we will see in Sect. 9.6. Thus, there are hidden depths in the seemingly elementary idea of congruence classes.

Dedekind's definition of an infinite set also involves countable AC, as we saw in the exercises to Sect. 7.1, since it involves choosing a countable infinity of members from an infinite set. Bettazzi (1896) was perhaps the first to question whether it is valid to make an infinite sequence of choices, but his objection was forgotten. It was not raised again until Zermelo (1904) made AC explicit and cited Dedekind's definition of infinite sets as an application.

AC crystallized in Zermelo's mind after an incident at the 1904 International Congress of Mathematicians in Heidelberg. At the Congress, in August, Julius König presented an argument that there is *no* well-ordering of $\mathbb{R}$. The next day, Zermelo found a mistake in König's reasoning, and began to work on the well-ordering problem himself. It became clear to him that AC was the key idea, and with it he obtained a proof of the well-ordering theorem on September 24, 1904. He sent it to Hilbert, who quickly arranged for its publication.

The reaction to Zermelo's proof was mostly hostile, even from mathematicians who had unconsciously used AC in their previous work. For example, Borel (1905) (writing on December 1, 1904) correctly identified the essence of Zermelo's proof, but denied that his argument could be part of mathematics:

> Such reasoning seems to me to be no more justified than the following: "To well order a set M, it suffices to choose arbitrarily an element to which one assigns rank 1, then another to which one assigns rank 2, and so on *transfinitely*; that is, until all elements of M have been exhausted ..." But no mathematician would regard this reasoning as valid.

> Borel (1905), p. 195.

Supporters of AC were fewer, and less eminent, yet they produced the two most lasting results of the immediate post-Zermelo years: the basis of $\mathbb{R}$ over $\mathbb{Q}$ due to Hamel (1905), and the nonmeasurable set of Vitali (1905).

As mentioned in the exercises to Sect. 7.2, Hamel used his basis to produce a discontinuous real function f with the *additive* property:

$$f(x + y) = f(x) + f(y).$$

The existence of such a function had been an open problem ever since Cauchy (1821), pp. 104–106, showed that the only *continuous* additive functions are those of the form $f(x) = ax$ for constant a. (Cauchy's proof was outlined in Exercises 3.4.5 and 3.4.6.) Cauchy's theorem has implications for the theory of measure on the line or the plane, where the measure $\mu(A \cup B)$ of disjoint sets A and B is supposed to be $\mu(A) + \mu(B)$. If μ varies continuously with the endpoints c and d of an interval on the line, then μ is necessarily a constant multiple of the usual length function $|d - c|$ on intervals. Similarly, a continuous measure on subsets of the plane is a constant multiple of the usual area function on rectangles.

Vitali (1905), as mentioned above, used the congruence classes of $\mathbb{R}$ to obtain a *non*measurable set. We defer a full explanation of measurability until Chap. 9, but it is worth mentioning here that the founders of measure theory, Borel and Lebesgue, rejected AC. They tended to accept countable AC (or to use it unconsciously), however, because measure theory is not really possible without it.

After a few years of hostility, support for AC began to grow. Steinitz (1910) used AC to prove that every field F has an *algebraic closure* $\overline{F}$; that is, $\overline{F} \supseteq F$ and every polynomial equation with coefficients in $\overline{F}$ has a solution in $\overline{F}$. This was the first of many results about "closed" or "maximal" algebraic structures that depend on AC, so algebraists became strong supporters of AC. Indeed, by the 1930s algebra was influencing the presentation of set theory by favoring maximal principles like Zorn's lemma[4] over the equivalent principles of AC and the well-ordering theorem. One can see this in the book of Bourbaki (1939) which does not use ordinals at all, and mentions them only in an exercise.

At the same time, set theorists investigated AC and discovered that it had many interesting and/or bizarre consequences. Some of the most interesting are nonmeasurable sets, but the *undetermined* sets of Sect. 7.6 are also of interest. The theorem on winning strategies for finite games that instigated this line of research is due to Zermelo (1913). The generalization to infinite games was entertained by the Polish mathematicians Steinhaus, Banach, and Mazur in the 1920s, as Steinhaus (1965) reported. But they dropped the idea after Banach and Mazur found that winning strategies do not always exist, if AC holds. The subject was revived in the 1960s by Steinhaus and Mycielski, who thought that AD could be a useful alternative to AC in analysis. Their confidence was borne out by Mycielski and Świerczkowski (1964), who proved that AD implies countable AC for sets of reals and that all subsets of $\mathbb{R}$ are measurable.

The consequences of AC were sometimes bizarre, but they were never contradictory, and Gödel (1938) explained why. As described in Sect. 7.3, he defined the class of constructible sets, which satisfies all the axioms of ZF *plus* AC. It follows that no contradiction can arise from AC unless ZF itself is contradictory.

So, the consequences of AC are not contradictory—but Gödel did not show that they are true. He could not, because Cohen (1963) showed that it is also consistent with ZF to assume that AC is *false*, even in some very specific instances. For example, Cohen showed that one can consistently assume that there is an infinite

---

[4]Zorn's lemma gets its name from Zorn (1935), but it is actually due to Kuratowski (1922). The name stuck after Bourbaki (1939) called it "Zorn's theorem."

Fig. 7.3 Kurt Gödel and Paul Cohen

set of real numbers with no countably infinite subset. Using Cohen's methods, Feferman and Levy (1963) showed that it is even consistent to assume that $\mathbb{R}$ is a countable union of countable sets.

The message of Gödel and Cohen's results is that the ZF axioms are very far from complete. They fail to settle many questions about the real numbers that are settled (often in contrary ways) by AC and AD. Thus, new axioms are called for, but so far none as compelling as the ZF axioms have been proposed. AC has been the most popular new axiom, because it makes the universe more orderly from at least two points of view. For algebraists, AC brings complete or maximal structures, such as the algebraic closure of any field; for set theorists, AC makes every set well-ordered, hence any two sets are comparable in cardinality. However, AC is disruptive at the level of $\mathbb{R}$, where it creates nonmeasurable and undetermined sets. At this level, AD has some advantages: it blocks nonmeasurable and undetermined sets, yet allows countable AC for sets of reals, which is enough AC for most of analysis.

7.9.1 AC, AD, and the Natural Numbers

Since AC and AD have a profound influence on the properties of the real numbers, one wonders whether they also affect theorems about the *natural* numbers. Could it be, for example, that every even number > 2 is the sum of two primes if AC holds, but not otherwise? Fortunately, no. If a theorem about natural numbers (involving only elementary concepts such as addition and multiplication) is provable with the help of AC, then it is also provable without AC.

The explanation of this fact lies in Gödel's class L of constructible sets, mentioned in Sect. 7.3. The natural numbers (and other elementary concepts) have the same meaning in L as they do in the universe of all sets, so proving that a theorem T about natural numbers holds in L amounts to proving T outright. But, as mentioned in Sect. 7.3, AC is provable in L, so by proving T in L we can avoid assuming AC. (In contrast, the real numbers do *not* have the same meaning in L as in the universe of all sets, because it is possible that nonconstructible reals exist. Indeed, the existence of nonconstructible reals follows from the presence of certain large sets, which are generally believed to exist though admittedly that is not provable in ZF.)

It is a similar, though more complicated, story for AD. If one applies Gödel's set construction operations to the set $\mathbb{R}$, one obtains the class $L(\mathbb{R})$ of sets "constructible from $\mathbb{R}$." $L(\mathbb{R})$ is again a model of the ZF axioms, and proving that a theorem T about natural numbers holds in $L(\mathbb{R})$ amounts to proving T outright. Indeed, $L(\mathbb{R})$ is just the same as L unless we assume the existence of sets large enough to imply the existence of nonconstructible reals. But if we assume the existence of sufficiently large sets a wonderful thing happens: AD *can be proved to hold in* $L(\mathbb{R})$. It then follows that AD can be eliminated from the proof of any theorem about the natural numbers.

The proof that AD holds in $L(\mathbb{R})$ is a very difficult one due to Woodin in 1985. Woodin's proof was never published, but a proof was included in Martin and Steel (1989), along with other deep results on determinacy. Neeman (2010) is a recent paper entirely dedicated to the proof that AD holds in $L(\mathbb{R})$, assuming that sufficiently large sets exist.

Chapter 8
Borel Sets

PREVIEW

The Borel sets may be described simply as those generated from the open sets by the operations of complementation and countable union. But one gains a clearer understanding of Borel sets by dividing their generation into stages numbered by countable ordinals, with Σ_α denoting the class of Borel sets generated by stage α.

The foundation for the classification of Borel sets is the universal open set constructed in Sect. 5.6. In this chapter we work with subsets of $\mathcal{N}$, the set of irrational numbers in [0,1], as we did in there.

It turns out that the construction of a universal open set "propagates" through the Borel sets to give a universal Σ_α set for each α, and the diagonal argument then shows that Σ_α includes Borel sets not in Σ_β for any $\beta < \alpha$. Thus, the Borel sets are arranged in a *hierarchy*, with new sets appearing continually as α increases.

The Borel sets do not exhaust all subsets of $\mathcal{N}$, since we can show that there are only $2^{\aleph_0}$ Borel sets—as many as there are members of $\mathcal{N}$ or $\mathbb{R}$. But they do show the scope of the countable union operation (important for the concept of *measure* explored in the next chapter), and the related operation of forming the limit of a sequence of functions. Indeed, the functions generated from continuous functions by taking limits also form a hierarchy, closely related to the Borel hierarchy, called the *Baire hierarchy*.

8.1 Borel Sets

The class $\mathcal{B}$ of Borel sets may be defined in two rather different ways: as the *closure* of the class of open sets under the operations of complement and countable union, or as the *union of a sequence* of classes (the first of which is the class of open sets) defined by transfinite induction. The "closure" definition is easier to state, so we consider it first.

J. Stillwell, *The Real Numbers: An Introduction to Set Theory and Analysis*,
Undergraduate Texts in Mathematics, DOI 10.1007/978-3-319-01577-4_8,
© Springer International Publishing Switzerland 2013

Definition 1. $\mathcal{B}$ is the least set that includes all open sets and is closed under complement and union. That is, $\mathcal{B}$ is the intersection of all sets $\subseteq \mathcal{P}(\mathcal{N})$ that include all open sets and are closed under complement and countable union.

To be more precise, $\mathcal{B}$ is the intersection of all sets $S \subseteq \mathcal{P}(\mathcal{N})$ with the properties

(i) Each open subset of $\mathcal{N}$ belongs to S.
(ii) If $X \in S$ then $\mathcal{N} - X \in S$.
(iii) If $X_1, X_2, X_3, \ldots \in S$ then $(X_1 \cup X_2 \cup X_3 \cup \cdots) \in S$.

Defining a set by closure properties, as here, is typical in modern mathematics. However, the glibness of this definition hides a property of Borel sets we would like to see: their *levels of complexity*. Open sets are naturally viewed as the simplest Borel sets, and other Borel sets have a complexity that can be measured by the number of complements and countable unions needed to construct them. This number can be an arbitrary countable ordinal, as we will see in Sect. 8.4. This brings us to the second definition of the class $\mathcal{B}$ of Borel sets, where ordinals make their appearance.

Definition 2. $\mathcal{B}$ is the union of the classes Σ_α defined inductively as follows (along with classes Π_α) for all countable ordinals α.

$$\Sigma_1 = \{\text{open subsets of } \mathcal{N}\}$$

$$\Pi_1 = \{\mathcal{N} - X : X \in \Sigma_1\}$$

$$\Sigma_\alpha = \left\{\text{countable unions of sets in } \bigcup_{\beta < \alpha} \Pi_\beta\right\}$$

$$\Pi_\alpha = \{\mathcal{N} - X : X \in \Sigma_\alpha\}.$$

It is not clear that all the countable ordinals are really needed in this definition, because it is not clear that each Σ_α includes sets not in any Σ_β for $\beta < \alpha$. This will be proved in Sect. 8.4.

Notice that Σ_α is defined in a way that avoids distinguishing between successor and limit ordinals α. This greatly assists some later constructions, though it makes the successor ordinal case look more complicated than necessary. In fact (see exercises below)

$$\Sigma_{\beta+1} = \{\text{countable unions of sets in } \Pi_\beta\}.$$

It is not particularly hard to show that Definition 1 and Definition 2 are equivalent, but the proof depends on countable AC. More specifically, it depends on the consequence of countable AC that a countable union of countable sets is countable.

Equivalent definitions of Borel sets. *Definitions 1 and 2 define the same class of sets.*

Proof. The class $\bigcup_{\alpha < \omega_1} \Sigma_\alpha$ from Definition 2 certainly includes the open sets, because they comprise the subset Σ_1. Next we show that $\bigcup_{\alpha < \omega_1} \Sigma_\alpha$ is closed under complementation and countable unions. First,

$$X \in \Sigma_\beta \Rightarrow \mathcal{N} - X \in \Pi_\beta \quad \text{by definition of } \Pi_\beta$$
$$\Rightarrow \mathcal{N} - X \in \Sigma_\alpha \quad \text{for any } \alpha > \beta,$$

since $\mathcal{N} - X$ is (trivially) a countable union of copies of itself. Thus, $\bigcup_{\alpha < \omega_1} \Sigma_\alpha$ is closed under complements.

Observe, as a by-product of this argument, that $X \in \Sigma_\beta$ implies $X \in \Pi_\alpha$ for any $\alpha > \beta$. Applying this observation to any $X_1, X_2, X_3, \ldots \in \bigcup_{\alpha < \omega_1} \Sigma_\alpha$, we find that each $X_i \in$ some Π_{α_i}. By countable AC, the union γ of the countably many countable sets $\alpha_1, \alpha_2, \alpha_3, \ldots$ is a countable ordinal, so

$$X_1 \cup X_2 \cup X_3 \cup \cdots \in \Sigma_{\gamma+1}$$

by definition of $\Sigma_{\gamma+1}$. This shows that $\bigcup_{\alpha < \omega_1} \Sigma_\alpha$ is closed under countable unions, and completes the proof that $\bigcup_{\alpha < \omega_1} \Sigma_\alpha$ has the closure properties stated in Definition 1.

To show that $\bigcup_{\alpha < \omega_1} \Sigma_\alpha$ is the *least* such set, it suffices to show that any set with these closure properties contains each Σ_α. This is immediate from the inductive Definition 2, since the operations used in the definition are complementation and countable union.   □

Exercises

8.1.1 Show that any countable subset of $\mathcal{N}$ is in Σ_2. Why does this show that Σ_2 has members not in Σ_1?

8.1.2 Prove by induction on α that $\Sigma_\alpha \subseteq \Sigma_{\alpha+1}$, $\Pi_\alpha \subseteq \Pi_{\alpha+1}$, $\Pi_\alpha \subseteq \Sigma_{\alpha+1}$, and $\Sigma_\alpha \subseteq \Pi_{\alpha+1}$.

8.1.3 Deduce that $\Sigma_{\alpha+1} = \{$countable unions of sets in $\Pi_\alpha\}$.

8.1.4 Prove that a countable union of Π_α sets is $\Sigma_{\alpha+1}$, and a countable intersection of Σ_α sets is $\Pi_{\alpha+1}$.

8.1.5 Let $\alpha_1 < \alpha_2 < \alpha_3 < \cdots$ be a fixed sequence of countable ordinals with limit λ. Show that any $X \in \Sigma_\lambda$ is of the form

$$X = X_1 \cup X_2 \cup X_3 \cup \cdots \quad \text{where } X_i \in \Pi_{\alpha_i}.$$

(*Hint*: Insert empty sets X_j in the sequence $X_1, X_2, X_3, \ldots$ where necessary.)

8.2 Borel Sets and Continuous Functions

In the previous section we defined the Borel subsets of $\mathcal{N}$. There are similar definitions for the Borel subsets of $\mathcal{N}^2, \mathcal{N}^3, \ldots$ and so on. The bottom level Σ_1 consists of the open subsets of $\mathcal{N}^k$, and these are again countable unions of *basic* open sets. The only difference is that a basic open set in $\mathcal{N}^k$ is the *cartesian product* of basic open sets in each factor $\mathcal{N}$. For example, a basic open subset of $\mathcal{N}^2$ is of the form $I \times J$, where I and J are basic open subsets of $\mathcal{N}$. $I \times J$ is a rectangle minus each point with a rational coordinate.

We are particularly interested in the Borel subsets of $\mathcal{N}^2$, because we wish to generalize the universal open set $\mathcal{U} \subset \mathcal{N}^2$ of Sect. 5.6 to a set $\mathcal{U}_\alpha \subset \mathcal{N}^2$ which is "universal Σ_α" in the same sense. That is, $\mathcal{U}_\alpha$ is a Σ_α subset of $\mathcal{N}^2$, and its sections

$$\mathcal{U}_\alpha(y) = \{x : \langle x, y \rangle \in \mathcal{U}_\alpha\}$$

are all the Σ_α subsets of $\mathcal{N}$.

To make this possible we generalize the basic property of continuous functions from open sets to Σ_α sets:

Inverse images of Borel sets under continuous functions. *For each countable ordinal α, and each continuous function f from $\mathcal{N}^j$ onto $\mathcal{N}^k$, f^{-1} of a Σ_α set $X \subseteq \mathcal{N}^k$ is Σ_α, and f^{-1} of a Π_α set $X \subseteq \mathcal{N}^k$ is Π_α.*

Proof. We argue by induction on α. For $\alpha = 1$, f^{-1} of a Σ_α set is f^{-1} of an open set, hence open; that is, Σ_1. It follows that f^{-1} of a Π_1 set is Π_1 by taking complements.

For the induction step, suppose f^{-1} of a Σ_β set is Σ_β, and f^{-1} of a Π_β set is Π_β, for all $\beta < \alpha$. By Definition 2 of Borel sets, $X \in \Sigma_\alpha$ is of the form

$$X = X_1 \cup X_2 \cup X_3 \cup \cdots,$$

where each of $X_1, X_2, X_3, \ldots$ belongs to some Π_β with $\beta < \alpha$. Therefore

$$f^{-1}(X) = f^{-1}(X_1) \cup f^{-1}(X_2) \cup f^{-1}(X_3) \cup \cdots,$$

and each term in the union is in some Π_β, with $\beta < \alpha$, by the induction hypothesis. This implies

$$f^{-1}(X) \in \Sigma_\alpha \quad \text{by Definition 2.}$$

It then follows, by taking complements again, that f^{-1} of a Π_α set is Π_α, and the induction is complete. $\square$

The second tool for the construction of a universal Σ_α set is a "continuous encoding" of each sequence of irrational numbers (members of $\mathcal{N}$) by a single irrational number. Given this tool, if we can encode Π_β sets, for $\beta < \alpha$, by members of $\mathcal{N}$, then we can also encode their countable unions (the Σ_α sets) by members of $\mathcal{N}$, in a continuous fashion.

Continuous bijection $g : \mathcal{N} \to \mathcal{N}^{\mathbb{N}}$. *There is a function g, sending each $y \in \mathcal{N}$ to a unique sequence $\langle y_1, y_2, y_3, \ldots \rangle \in \mathcal{N}^{\mathbb{N}}$, such that each member of $\mathcal{N}^{\mathbb{N}}$ is a value $g(y)$, and each of the coordinate functions $g_k(y) = y_k$ is continuous.*

Proof. If $y = \langle a_1, a_2, a_3, a_4, a_5, \ldots \rangle$, then one way to define $g_1, g_2, g_3, \ldots$ is the following:

$$g_1(y) = \langle a_1, a_3, a_5, a_7, \ldots \rangle,$$

$$g_2(y) = \langle a_2, a_6, a_{10}, a_{14}, \ldots \rangle,$$

$$g_3(y) = \langle a_4, a_{12}, a_{20}, a_{28}, \ldots \rangle,$$

$$\vdots$$

In other words, $g_1(y)$ omits every other term of the sequence y; $g_2(y)$ omits every other term of the sequence omitted by g_1; $g_3(y)$ omits every other term of the sequence omitted by g_1 and g_2; and so on.

It is clear that the sequence

$$g(y) = \langle y_1, y_2, y_3, \ldots \rangle = \langle g_1(y), g_2(y), g_3(y) \ldots \rangle$$

is uniquely determined by y. Also, any sequence $\langle y_1, y_2, y_3, \ldots \rangle$ in $\mathcal{N}^{\mathbb{N}}$ is obtainable for suitable choice of y, because y can in fact be assembled from $\langle y_1, y_2, y_3, \ldots \rangle$. Thus, $g : \mathcal{N} \to \mathcal{N}^{\mathbb{N}}$ is a bijection.

Finally, each g_k is continuous. Because we can make $g_k(y')$ arbitrarily close to $g_k(y)$ by choosing y' sufficiently close to y; that is, by making y' agree with y on a sufficiently long initial segment $\langle a_1, a_2, a_3, \ldots, a_k \rangle$. □

This proof works because y is separated (or *partitioned*) into infinitely many disjoint subsequences. Any other partition of y works equally well: we still get a bijection $y \leftrightarrow \langle y_1, y_2, y_3, \ldots \rangle$, and $g_k(y')$ is arbitrarily close to $g_k(y)$ if y' and y agree on a sufficiently long initial segment. Thus, y in a reasonable sense *encodes* the sequence $\langle y_1, y_2, y_3, \ldots \rangle$, and the encoding is continuous.

Exercises

8.2.1 Use the bijection $p : \mathbb{N}^2 \to \mathbb{N}$ from Sect. 3.2 to obtain another bijection $\mathcal{N} \to \mathcal{N}^{\mathbb{N}}$.

8.2.2 Show that the function $g : \mathcal{N} \to \mathcal{N}^{\mathbb{N}}$ has a unique continuous extension $\bar{g} : [0, 1] \to [0, 1]^{\mathbb{N}}$.

8.2.3 Is $\bar{g}$ a surjection? A bijection?

8.3   Universal Σ_α Sets

With the two theorems of the previous section—preservation of the Σ_α and Π_α properties by (inverses of) continuous functions, and the continuous encoding of sequences by single elements of $\mathcal{N}$—we are now ready to extend the construction of a universal set from Σ_1 to all levels of the Borel hierarchy.

Universal Σ_α set. *For each countable ordinal α there is a Σ_α set $\mathcal{U}_\alpha \subset \mathcal{N}^2$ whose sections*

$$\mathcal{U}_\alpha(y) = \{x : \langle x, y \rangle \in \mathcal{U}_\alpha\}$$

are all the Σ_α subsets of $\mathcal{N}$.

Proof. We argue by induction on α. For $\alpha = 1$ we can take $\mathcal{U}_1$ to be the universal open set constructed in Sect. 5.6, since open sets are Σ_1.

For the induction step we suppose that for each $\beta < \alpha$ there is a Σ_β set $\mathcal{U}_\beta$ whose sections

$$\mathcal{U}_\beta(y) = \{x : \langle x, y \rangle \in \mathcal{U}_\beta\}$$

are all the Σ_β subsets of $\mathcal{N}$. (More precisely, we have to *choose* a universal Σ_β set $\mathcal{U}_\beta$ for each of the countably many $\beta < \alpha$, using countable AC.) It follows that $\mathcal{N}^2 - \mathcal{U}_\beta$ is a universal Π_β set, because its sections are precisely the complements of the sections $\mathcal{U}_\beta(y)$; that is, the Π_β subsets of $\mathcal{N}$.

Now to form Σ_α sets we need to form countable unions of Π_β sets for (possibly various) $\beta < \alpha$. To do this uniformly we first choose a sequence $\beta_1 \leq \beta_2 \leq \beta_3 \leq \cdots$ with limit α if α is a limit ordinal and limit γ if $\alpha = \gamma + 1$. In either case, each $X \in \Sigma_\alpha$ is of the form

$$X = X_1 \cup X_2 \cup X_3 \cup \cdots \quad \text{where each } X_n \in \Pi_{\beta_n}.$$

By our induction hypothesis we have a universal Σ_{β_n} set $\mathcal{U}_{\beta_n}$ for each β_n, of which X_n is the complement of a section $\mathcal{U}_{\beta_n}(y_n)$. That is,

$$x \in X_n \Leftrightarrow \langle x, y_n \rangle \notin \mathcal{U}_{\beta_n}.$$

And therefore

$$x \in X \Leftrightarrow \langle x, y_n \rangle \notin \mathcal{U}_{\beta_n} \quad \text{for some } n.$$

We also know, from the previous section, that each sequence $\langle y_1, y_2, y_3, \ldots \rangle$ occurs as $\langle g_1(y), g_2(y), g_3(y), \ldots \rangle$ for some $y \in \mathcal{N}$. For this y we therefore have

$$x \in X \Leftrightarrow \langle x, g_n(y) \rangle \notin \mathcal{U}_{\beta_n} \quad \text{for some } n.$$

In this sense y "encodes" the Σ_α set X, and we have a "universal" set $\mathcal{U}_\alpha$ defined by

$$\langle x, y \rangle \in \mathcal{U}_\alpha \Leftrightarrow \langle x, g_n(y) \rangle \notin \mathcal{U}_{\beta_n} \quad \text{for some } n,$$

because all of its sections $\mathcal{U}_\alpha(y) = \{x : \langle x, g_n(y) \rangle \notin \mathcal{U}_{\beta_n} \text{ for some } n\}$ are Σ_α, and they include all the Σ_α sets X. Thus, it remains only to prove that $\mathcal{U}_\alpha$ is itself a Σ_α set.

It suffices to show that $\mathcal{U}_\alpha$ is a countable union of $\mathbf{\Pi}_{\beta_n}$ sets, and indeed that, for each n,

$$\{\langle x, y \rangle : \langle x, g_n(y) \rangle \notin \mathcal{U}_{\beta_n}\} \quad \text{is} \quad \mathbf{\Pi}_{\beta_n}.$$

We show, equivalently, that its complement

$$\mathcal{V}_n = \{\langle x, y \rangle : \langle x, g_n(y) \rangle \in \mathcal{U}_{\beta_n}\} \quad \text{is} \quad \mathbf{\Sigma}_{\beta_n}.$$

To do this we recall that g_n is continuous, and hence so is the function $\langle x, y \rangle \mapsto \langle x, g_n(y) \rangle$. $\mathcal{V}_n$ is the inverse image of the $\mathbf{\Sigma}_{\beta_n}$ set $\mathcal{U}_{\beta_n}$ under the latter function. So $\mathcal{V}_n$ is also $\mathbf{\Sigma}_{\beta_n}$, as required, by the first theorem of the previous section. $\qquad \square$

Exercises

8.3.1 Explain why the sets $\mathcal{U}_\alpha(y) = \{x : \langle x, g_n(y) \rangle \notin \mathcal{U}_{\beta_n} \text{ for some } n\}$ are $\mathbf{\Sigma}_\alpha$.

8.3.2 Re-prove the induction step in the special case where $\alpha = \gamma + 1$. (In this case we can assume that $X \in \mathbf{\Sigma}_{\gamma+1}$ has the form $X = X_1 \cup X_2 \cup X_3 \cup \cdots$ where each $X_n \in \mathbf{\Pi}_\gamma$, so there is no need to use a sequence $\beta_1 \leq \beta_2 \leq \beta_3 \leq \cdots$.)

As noted in the second paragraph of the proof, we are using countable AC. Indeed we cannot prove that the Borel hierarchy extends beyond $\mathbf{\Sigma}_3$ without using countable AC. This is due to the result, mentioned in the exercises to Sect. 7.7, that it is consistent with ZF to assume that $\mathbb{R}$ is a countable union of countable sets.

8.3.3 Under this assumption, show that all sets of real numbers are in $\mathbf{\Sigma}_3$.

8.4 The Borel Hierarchy

We are now ready to show that each level $\mathbf{\Sigma}_\alpha$ of the Borel hierarchy has members not in $\mathbf{\Sigma}_\beta$ for any $\beta < \alpha$. We know that

1. $\mathbf{\Sigma}_\alpha$ includes each member X of $\mathbf{\Pi}_\beta$ for any $\beta < \alpha$ (as the union of countably many copies of X).

So if we can prove that

2. $\mathbf{\Pi}_\alpha$ has a member Y *not* in $\mathbf{\Sigma}_\alpha$,

then it will follow that

3. $\mathbf{\Pi}_\alpha$ has a member Y not in $\mathbf{\Pi}_\beta$ for any $\beta < \alpha$.

Therefore (taking complements),

4. $\mathbf{\Sigma}_\alpha$ has a member $\mathcal{N} - Y$ not in $\mathbf{\Sigma}_\beta$ for any $\beta < \alpha$.

Borel hierarchy theorem. Π_α *has a member not in* Σ_α *(and therefore, as explained above,* Σ_α *has a member not in* Σ_β *for any* $\beta < \alpha$*).*

Proof. We apply the diagonal argument to the universal Σ_α set $\mathcal{U}_\alpha$ from the previous section. By construction, the sections of $\mathcal{U}_\alpha$, $\mathcal{U}_\alpha(y) = \{x : \langle x, y \rangle \in \mathcal{U}_\alpha\}$, are all the Σ_α subsets of $\mathcal{N}$. But the "complementary diagonal set" $\mathcal{D}_\alpha = \{x : \langle x, x \rangle \notin \mathcal{U}_\alpha\}$ is unequal to each section $\mathcal{U}_\alpha(y)$. In fact,

$$y \in \mathcal{D}_\alpha \Leftrightarrow \langle y, y \rangle \notin \mathcal{U}_\alpha \Leftrightarrow y \notin \mathcal{U}_\alpha(y),$$

so $\mathcal{D}_\alpha$ differs from $\mathcal{U}_\alpha(y)$ regarding the element y. Thus, $\mathcal{D}_\alpha$ is not in Σ_α.

But $\mathcal{D}_\alpha = \{x : \langle x, x \rangle \notin \mathcal{U}_\alpha\}$ is in Π_α, as we prove by induction on α.

For the base step, $\alpha = 1$, we observe that the open set $\mathcal{U}_1$ is a countable union of open rectangles, so it meets the diagonal $\{\langle x, x \rangle : x \in \mathcal{N}\}$ in a countable union of open intervals. Then its projection $\{x : \langle x, x \rangle \in \mathcal{U}_1\}$ on the x-axis is also a countable union of intervals, hence Σ_1. The complement $\mathcal{D}_1$ of this projection is therefore Π_1.

Notice that this argument applies with any Σ_1 set $\mathcal{V}_1$ in place of $\mathcal{U}_1$.

For the induction step our hypothesis is that, for each $\beta < \alpha$,

$$\{x : \langle x, x \rangle \notin \mathcal{V}_\beta\} \quad \text{is } \Pi_\beta \text{ for any } \Sigma_\beta \text{ set } \mathcal{V}_\beta, \quad \text{or equivalently}$$

$$\{x : \langle x, x \rangle \in \mathcal{W}_\beta\} \quad \text{is } \Pi_\beta \text{ for any } \Pi_\beta \text{ set } \mathcal{W}_\beta.$$

Now consider $\mathcal{D}_\alpha = \{x : \langle x, x \rangle \notin \mathcal{U}_\alpha\}$. We know $\mathcal{U}_\alpha = \mathcal{W}_1 \cup \mathcal{W}_2 \cup \cdots$, where each $\mathcal{W}_i \in \Pi_{\beta_i}$ for some $\beta_i < \alpha$, by definition of Σ_α. So

$$\mathcal{D}_\alpha = \{x : \langle x, x \rangle \notin \mathcal{U}_\alpha\} = \mathcal{N} - \{x : \langle x, x \rangle \in \mathcal{U}_\alpha\}$$

$$= \mathcal{N} - \{x : \langle x, x \rangle \in \bigcup_{i=1}^\infty \mathcal{W}_i\}$$

$$= \mathcal{N} - \bigcup_{i=1}^\infty \{x : \langle x, x \rangle \in \mathcal{W}_i\}$$

$$= \mathcal{N} - \bigcup_{i=1}^\infty (\text{some } \Pi_{\beta_i} \text{ set}) \quad \text{by induction}$$

$$= \mathcal{N} - (\text{some } \Sigma_\alpha \text{ set}) \quad \text{since } \beta_i < \alpha,$$

which is in Π_α. This completes the induction, so $\mathcal{D}_\alpha = \{x : \langle x, x \rangle \notin \mathcal{U}_\alpha\}$ is a Π_α set not in Σ_α, as required.   □

The Borel hierarchy theorem gives content to the following definition, which captures the concept of "complexity" of Borel sets S.

Definition. The least α such that $S \in \Sigma_\alpha$ is called the *Borel rank* of S.

By the hierarchy theorem, all countable ordinals occur as Borel ranks.

Exercises

Most "natural" examples of Borel sets occur at quite low levels of the hierarchy. An example is the set of *normal numbers*, where a number x is called *normal* (in base 2) if the digits 0 and 1 occur with equal frequency in the binary expansion of x. In other words: if there are m occurrences of the digit 1 in the first n binary digits of x, then $m/n \to 1/2$ as $n \to \infty$.

8.4.1 By formalizing the above statement about limits, show that

$$x \text{ is normal} \Leftrightarrow \text{for all } \varepsilon > 0 \text{ there is an } N \text{ such that}$$

$$\frac{1}{2} - \varepsilon \le \frac{m}{n} \le \frac{1}{2} + \varepsilon \quad \text{for all } n > N,$$

where m = number of occurrences of 1 in the first n digits of x.

Now we translate this statement about digits into a statement about sets.

8.4.2 Explain why, if $[0,1]$ is divided into 2^n equal subintervals I_k, the numbers x in $I_k \cap N$ agree in their first n binary digits. (So it makes sense to speak of "the number of occurrences, m, of 1 in the first n digits" in I_k.)

8.4.3 For each $\varepsilon > 0$ and positive integer n let $U_{\varepsilon,n}$ be the union of the intervals I_k for which

$$\frac{1}{2} - \varepsilon \le \frac{m}{n} \le \frac{1}{2} + \varepsilon.$$

Explain why $U_{\varepsilon,n}$ is a Σ_1 set.

8.4.4 Explain why

$$x \text{ is normal} \Leftrightarrow \text{for all } \varepsilon, \; x \in \bigcup_N \bigcap_{n>N} U_{\varepsilon,n}.$$

Finally, we restrict the values of ε to $\varepsilon = 1/M$, for positive integers M.

8.4.5 Explain why

$$x \text{ is normal} \Leftrightarrow x \in \bigcap_M \bigcup_N \bigcap_{n>N} U_{1/M,n},$$

and deduce that the set of normal numbers is Π_4. (A more refined argument actually shows that the set is Π_3.)

8.5 Baire Functions

After this long trek into the wilderness of Borel sets, we are in a better position to understand the interaction between two basic concepts of analysis: continuity and limits. We know from Sect. 4.6 that a *uniformly* convergent sequence of continuous functions has a continuous limit, so uniformly convergent sequences lead to nothing new. But, as we also saw in Sect. 4.6, a merely convergent sequence of continuous functions may have a discontinuous limit. An example is the function

$$f(x) = \begin{cases} 1 \text{ if } x = 0 \\ 0 \text{ if } x \neq 0, \end{cases}$$

which is the limit of continuous functions with a "spike" at $x = 0$ (see the picture in Sect. 4.6).

By repeatedly taking limits of previously defined functions we obtain an increasing sequence of function classes B_α called the *Baire hierarchy*. The classes B_α are defined for all countable ordinals α as follows.

$B_0 = \{\text{continuous functions } \mathbb{R} \to \mathbb{R}\}$

$B_\alpha = \{\text{limits of convergent sequences of functions from the } B_\beta \text{ with } \beta < \alpha.\}$

In this section we will only sketch some basic results on Baire functions, since we do not need the results later. For this purpose we can stick with the domain $\mathbb{R}$, since the domain $\mathcal{N}$ becomes convenient (as with Borel sets) only when more technical details are required. In $\mathbb{R}$ we have the rational numbers, which are involved in some of the most interesting Baire functions in the low levels of the hierarchy.

For example, the discontinuous function $f(x)$ defined above is a member of B_1, and so is the function $f(x, r)$ defined for each rational number r by

$$f(x, r) = \begin{cases} 1 \text{ if } x = r \\ 0 \text{ if } x \neq r. \end{cases}$$

If we then take an enumeration $r_1, r_2, r_3, \ldots$ of the rational numbers the functions

$$g_n(x) = f(x, r_1) + \cdots + f(x, r_n)$$

are also in B_1, because each g_n is the limit of a sequence of continuous functions with "spikes" at $r_1, \ldots, r_n$. The limit of the sequence $g_1, g_2, g_3, \ldots$ exists and equals

$$g(x) = \begin{cases} 1 \text{ if } x \text{ is rational} \\ 0 \text{ if } x \text{ is irrational.} \end{cases}$$

Thus, the highly discontinuous *Dirichlet function* of Sect. 1.5 is in B_2. This is what we meant when we said that the Dirichlet function is "not far removed" from continuity.

It is easy to guess that there is a connection between Baire functions and Borel sets, and the connection becomes clearer when continuity and limits are expressed in terms of sets. The key facts are the following.

1. As we already know from Sect. 5.2, a function f is continuous if and only if $f^{-1}(U)$ is open for any open set U. Thus, f^{-1} of a Σ_1 set is Σ_1, when f is continuous.
2. If $f(x) = \lim_{n \to \infty} f_n(x)$ and U is open, then

$$x \in f^{-1}(U) \Leftrightarrow f(x) \in U$$

$$\Leftrightarrow f_n(x) \in U \text{ for all sufficiently large } n$$

$$\Leftrightarrow \text{there is an } m \text{ such that } f_n(x) \in U \text{ for all } n > m$$

$$\Leftrightarrow \text{there is an } m \text{ such that } x \in f_n^{-1}(U) \text{ for all } n > m$$

$$\Leftrightarrow \text{there is an } m \text{ such that } x \in \bigcap_{n>m} f_n^{-1}(U)$$

$$\Leftrightarrow x \in \bigcup_{m} \bigcap_{n>m} f_n^{-1}(U).$$

Thus, $f^{-1}(U)$ is a countable union of countable intersections of the sets $f_n^{-1}(U)$.

These two facts enable an inductive proof that *for any Baire function f, f^{-1} of an open set is in Σ_β, for some β*. A more careful proof (see, e.g., Kechris (1995), p. 190) shows in fact that f^{-1} of an open set is in $\Sigma_{\alpha+1}$ when f is in B_α.

Conversely, one can show that every Borel set arises as f^{-1} of an open set for some Baire function f. In fact, one can show that the *characteristic function* χ_A of each Borel set A is Baire, where

$$\chi_A(x) = \begin{cases} 1 \text{ if } x \in A \\ 0 \text{ if } x \notin A. \end{cases}$$

For example, the characteristic function of the Σ_2 set $\mathbb{Q}$ is the Dirichlet function, which we have shown to be in B_2. To prove that the characteristic function of any Borel set is Baire, we use induction on the construction of Borel sets. The base step is to show that the characteristic function of any open interval (a, b) is in B_1 (exercise).

For the induction step, we use two closure properties of Baire functions that are easy to prove (see exercises):

Closure Under Sums.   If f and g are Baire then so is $f + g$.
Closure Under Composites.   If f and g are Baire, then so is $f \circ g$, defined by $(f \circ g)(x) = f(g(x))$.

Assuming that these closure properties hold, first suppose that A is a Borel set whose complement $\mathcal{N} - A$ has a Baire characteristic function. Then χ_A is the composite of $\chi_{\mathcal{N}-A}$ with the continuous function

$$g(x) = 1 - x,$$

which exchanges the values 0 and 1. The function g is certainly Baire, hence so is χ_A.

Finally, suppose that A is the union

$$A = A_1 \cup A_2 \cup A_3 \cup \cdots,$$

where each of the characteristic functions $\chi_{A_1}, \chi_{A_2}, \chi_{A_3}, \ldots$ is Baire. By closure under sums, each of the functions

$$f_n(x) = \chi_{A_1}(x) + \cdots + \chi_{A_n}(x)$$

is also Baire. Clearly,

$$f_n(x) = \begin{cases} \text{some positive integer if } x \in A_1 \cup \cdots \cup A_n \\ 0 \qquad\qquad\qquad\qquad\qquad\text{otherwise.} \end{cases}$$

So if we compose f_n with the Baire function

$$h(x) = \begin{cases} 1 \text{ if } x > 0 \\ 0 \text{ otherwise,} \end{cases}$$

then we get $\chi_{A_1 \cup \cdots \cup A_n}$. Finally, χ_A is Baire, as the limit of the Baire functions $\chi_{A_1 \cup \cdots \cup A_n}$.

Thus, we have an inductive proof that χ_A is Baire for any Borel set A, and hence that every Borel set has the form $f^{-1}(U)$ for some open set U and Baire function f (taking U to be a small open set that includes 1).

Combining this result with the previous result that $f \in B_\alpha$ gives Borel sets $f^{-1}(U)$ of bounded Borel rank (in fact, in $\Sigma_{\alpha+1}$), we conclude that no B_α includes all Baire functions. Thus, new Baire functions occur in B_α for arbitrarily large values of α, which means that the Baire classes B_α form a true hierarchy, like the Borel hierarchy.

Exercises

8.5.1 Show that $\bigcup_{\alpha < \omega_1} B_\alpha$ is the least class of functions that includes the continuous functions and is closed under limits.

8.5.2 Show by induction on α that if $f, g \in B_\alpha$ then $f + g \in B_\alpha$ and $f \circ g \in B_\alpha$.

8.5.3 Show that the characteristic function of an interval (a, b) is the limit of continuous functions.

8.5.4 Show that any open set is a countable union of disjoint intervals, so that its characteristic function is also a limit of continuous functions.

8.5.5 Explain why the following formula defines the Dirichlet function

$$\lim_{m \to \infty} \lim_{n \to \infty} ((\cos m! \pi x)^{2n}).$$

8.5.6 Use Exercise 8.5.5 to give an immediate proof that the Dirichlet function is in Baire class 2.

8.5.7 The formula of Pringsheim (1899), p. 7, has n in place of $2n$. Why is his formula also valid?

8.6   The Number of Borel Sets

There are at least $2^{\aleph_0}$ Borel sets, because there are that many irrational numbers in $\mathcal{N}$, and hence that many open intervals in $\mathcal{N}$. It follows that there are $2^{\aleph_0}$ sets in each Σ_α, because these are the sets $\mathcal{U}_\alpha(y)$, as y varies over $\mathcal{N}$. The encoding of Σ_α

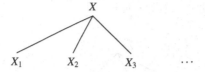

Fig. 8.1 Top portion of the tree for a Borel set X

sets by y values shows that there are at most $2^{\aleph_0}$ sets in Σ_α, and there are exactly this many because Σ_α includes all the open intervals. However, there are ω_1 values of α, so it remains unclear how many sets there are in $\bigcup_{\alpha<\omega_1} \Sigma_\alpha$.

There are in fact $2^{\aleph_0}$ Borel sets in total, and to show this we consider all Borel sets at once, encoding each of them by a *tree*.

A Borel set $X \in \Sigma_\alpha$ is naturally described by a tree, the top vertex of which represents X. Connected to this top vertex are vertices representing the sets $X_i \in \Pi_{\beta_i}$, where $\beta_i < \alpha$, such that

$$X = X_1 \cup X_2 \cup X_3 \cup \cdots .$$

Thus, the top two levels of the tree are as shown in Fig. 8.1. We then connect the vertex for $X_i \in \Pi_{\beta_i}$ to a vertex below it for $\mathcal{N} - X_i \in \Sigma_{\beta_i}$, and the latter vertex to vertices for the countable many sets (from Π_{γ_j} with $\gamma_j < \beta_i$) whose union is $\mathcal{N} - X_i$, and so on. Each branch of the tree corresponds to a descending sequence of ordinals $\alpha > \beta_i > \gamma_j > \cdots$, and hence *each branch is finite*. Moreover, each branch terminates in a basic open set G_n, which we can encode by the natural number n.

The set X is determined by the shape of its tree and the natural numbers n associated with its terminal vertices. Thus, the problem of encoding Borel sets amounts to describing the shapes of trees with finite branches and at most countably many descendants of each vertex, and labelling their terminal vertices with natural numbers. The means to do this are close at hand; namely, finite sequences $\langle a_1, a_2, \ldots, a_k \rangle$ of natural numbers.

We simply interpret $\langle a_1, a_2, \ldots, a_k \rangle$ as the branch with successive vertices $a_1, a_2, \ldots, a_k$, where a_k is the label on the terminal vertex and the other vertices are labelled in any way that correctly describes the shape of the tree. For example, it would be natural to label the top vertex by 1, and the vertices immediately below it by $1, 2, 3, \ldots$ (if these vertices are not terminal). It follows that *any Borel set can be encoded by a subset of the set* $\mathbb{N}^{<\omega}$ *of all finite sequences of natural numbers*. We saw in Sect. 3.1 (Example 8) that $\mathbb{N}^{<\omega}$ is countable, so there are as many Borel sets as there are subsets of a countable set, namely $2^{\aleph_0}$.

Exercises

8.6.1 Draw the tree for the open set $G_2 \cup G_4 \cup G_6 \cup \cdots$ and describe it by a set of ordered pairs of natural numbers.

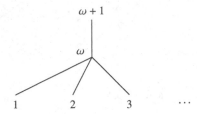

Fig. 8.2 Tree representing $\omega + 1$

Trees with branches of finite length and vertices of countable degree are also a useful way to encode countable ordinals. Each vertex is labelled by a countable ordinal α which (if the vertex is not terminal) is the least ordinal greater than the labels on the vertices below it. The terminal vertices are labelled by finite ordinals.

For example, Fig. 8.2 shows a tree representing the ordinal $\omega + 1$.

8.6.2 Draw a tree representing the ordinal $\omega \cdot 2$.

8.6.3 Prove by induction that every countable ordinal is representable by a tree with branches of finite length and vertices of countable degree.

8.7   Historical Remarks

The Borel sets get their name because of Borel (1898), where they are sketchily introduced on pp. 46–47 as a class of sets for which the concept of *measure* is meaningful. Given that the values of a measure are non-negative real numbers, and because we can form differences and infinite sums of real numbers, there are two arithmetic conditions that measurable subsets of [0,1] should satisfy:

Subtractivity.   If E and E' are measurable, with measures $\mu(E)$ and $\mu(E')$, and if $E' \subseteq E$, then

$$\mu(E - E') = \mu(E) - \mu(E').$$

Countable Additivity.   If $S_1, S_2, S_3, \ldots$ are disjoint sets with the measures $\mu(S_1)$, $\mu(S_2), \mu(S_3), \ldots$, then

$$\mu(S_1 \cup S_2 \cup S_3 \cup \cdots) = \mu(S_1) + \mu(S_2) + \mu(S_3) + \cdots.$$

Thus, implicitly, Borel is considering a class of sets $S \subseteq [0, 1]$ that is closed under the operations of difference and countable disjoint union. He does not mention the "base" sets of this class, but presumably they are sets whose measure is obvious, such as open or closed intervals.

It so happens that the Borel sets in [0,1] *can* be generated from intervals by differences and countable disjoint unions. However, this result is not obvious and was apparently first published by Sierpiński (1927). The present definition of Borel

Fig. 8.3 Henri Lebesgue

Fig. 8.4 René Baire and
Emile Borel

sets, using unrestricted countable unions, seems to have been used since 1902. This
was when the concept of measure reached its mature form, in the thesis of Lebesgue
(1902). Here Lebesgue introduced the concept of *Lebesgue measure* and proved its
basic properties, including countable additivity. We will say more about Lebesgue
measure, and why its scope includes all Borel sets, in Chap. 9.

The Baire classes of functions were introduced by Baire (1899), who observed
that the Dirichlet function is in Baire class 2. He did not investigate higher levels
of the Baire hierarchy, let alone prove that new functions appear at each level. This
was first done by Lebesgue (1905), who also proved that new sets appear at each
level of the Borel hierarchy.

It is noteworthy that Borel, Baire, and Lebesgue were all skeptical about AC,
even though they unconsciously used it (or at least countable AC) in their work.
They were among a large group of mathematicians who became painfully aware
of AC in 1904, after Zermelo published his proof of the well-ordering theorem.

Moore (1982) has an interesting account of this episode, including translations of 1905 correspondence between Borel, Baire, Lebesgue, and Hadamard (the "elder statesman" of the group and the only one to support AC).

Borel, Baire, and Lebesgue were interested in Borel sets and Baire functions because these objects are "definable by formulas" in some sense, and hence linked to classical analysis. We have seen for example, that the Dirichlet function is definable by the formula

$$\lim_{m \to \infty} \lim_{n \to \infty} ((\cos m! \pi x)^{2n}).$$

They were opposed to AC because it produces sets and functions with no apparent definitions. But with the realization that many results about Borel sets and Baire functions depend on countable AC came the realization that "definability" is not an absolutely clear concept. In retrospect, this was not surprising, because their "definitions" included series with infinitely many real number coefficients. In fact, "definability" is best confined to finite formulas and it is always a relative notion: one can speak of definability only in a given language, the symbols and syntax of which must be completely specified. Such *formal languages* were not used in mathematics until the 1920s. One such language is for ZF, where all definitions of functions must be available in order to state the replacement schema.

The discovery of the hierarchical properties of Borel sets by Lebesgue (1905) revealed, for the first time, a well-defined notion of *complexity* for sets of reals; namely, Borel rank. Set theorists could think about extending results known for open or closed sets to more complex sets in a systematic way. One such result was Cantor's theorem that every closed set F has the *perfect set property*: if uncountable, F contains a perfect subset (in which case F has the same cardinality as $\mathbb{R}$).

Hausdorff (1914), p. 466, mentioned that the perfect set property holds for sets with Borel rank ≤ 4, and remarked that there seemed hope of proving it for all Borel sets. As mentioned in Sect. 5.7, he did exactly that in Hausdorff (1916), and the same result was proved independently by Alexandrov (1916). From this theorem of Hausdorff and Alexandrov it follows that the continuum hypothesis holds for all Borel sets: if uncountable, a Borel set has the same cardinality as $\mathbb{R}$.

A rather similar story unfolded between the 1950s and the 1970s, with the concept of *determinacy* (the existence of a winning strategy) in place of the perfect set property. Gale and Stewart (1953) proved that open sets are determined, and determinacy for sets of Borel rank 2, 3, 4 was laboriously established over the next two decades. Then Martin (1975) proved that *all* Borel sets are determined. His proof was a tour de force, using the full resources of ZF+AC. In fact, Borel determinacy was known to be difficult before Martin proved it, because Friedman (1971) had shown that it is not provable in Zermelo set theory (which has AC but not the replacement schema). To this day, Borel determinacy is probably the best example of a theorem about the real numbers that depends on the replacement schema.

Nevertheless, Borel, Baire, and Lebesgue were right, in some sense, that the Borel sets are the sets most accessible to the human mind. They are the largest class of sets for which we can prove in ZF+AC all the "nice" properties: measurability, continuum hypothesis, and determinacy. In particular, determinacy is not provable in ZF+AC for the simplest natural extension of the Borel sets—the so-called *analytic* sets, which are the projections on $\mathbb{R}$ of Borel subsets of $\mathbb{R}^2$.

Chapter 9
Measure Theory

PREVIEW

In Sect. 1.7 we observed that any countable set has measure zero, because we can cover its first, second, third, ...points by intervals of lengths $\varepsilon/2, \varepsilon/4, \varepsilon/8, \ldots$. So the whole set can be covered by intervals of total length at most ε, which can be as small as we please.

This tells us that *countable sets can be ignored* in measure theory, but also that *countable union is a useful operation* for finding the measure of sets. In this chapter we will exploit countable unions of intervals to simultaneously define measure and to show that any measurable set can be approximated, within measure ε, by a finite union of intervals.

The concept of measure thus obtained is called *Lebesgue measure*, and it may be used to define a new concept of integral—the Lebesgue integral—that greatly extends the reach of classical calculus. Moreover, Lebesgue measure also clarifies the nature of the classical Riemann integral, by giving an exact description of the Riemann-integrable functions.

The scope of Lebesgue measure is so wide that the *non*measurable sets can be proved to exist only with the help of fairly strong forms of the axiom of choice. We give two examples. Because measurability of all sets of reals is incompatible with the full axiom of choice, it is not clear what set theory axioms are "ideal" for analysis. However, there are some interesting options, and in the Historical Remarks we discuss the set theory issues that they raise.

9.1 Measure of Open Sets

Since countable sets can be ignored in measure theory, it makes no difference whether we work with [0,1] (as contemplated in Sect. 1.7) or with the set of irrational numbers in [0,1]. The latter set can be identified with the set $\mathcal{N} = \mathbb{N}^{\mathbb{N}}$ of sequences of natural numbers, as we saw in Sect. 5.6. In the present chapter we will work with [0,1], because it and its subintervals are the simplest imaginable measurable sets, yet they suffice as a foundation for measuring highly complex sets, including all Borel sets.

The theory of Borel subsets of $\mathcal{N}$, developed in Chap. 8, transfers easily to [0,1]. The *open* subsets of [0,1] are the unions of open intervals, which now include the half-open "end intervals" of the form $[0, b)$ and $(a, 1]$. The Borel subsets of [0,1] are those subsets obtainable from open subsets by complementation and countable union. They are very similar to the Borel subsets of $\mathcal{N}$, from which they differ only by the possible presence of rational members.

In this section we show that each open set $U \subseteq [0, 1]$ has a measure $\mu(U)$ compatible with the usual length measure of closed intervals defined by

$$\mu([a, b]) = b - a.$$

We extend this measure μ to other intervals (open or half-open) and to countable unions of intervals by the following two rules:

Subtractivity   If $T \subseteq S$, and the sets S, T have measures $\mu(S), \mu(T)$ respectively, then $\mu(S - T) = \mu(S) - \mu(T)$.

Countable Additivity.   If $S_1, S_2, S_3, \ldots$ are disjoint sets with measures $\mu(S_1), \mu(S_2), \mu(S_3), \ldots$, then

$$\mu(S_1 \cup S_2 \cup S_3 \cup \cdots) = \mu(S_1) + \mu(S_2) + \mu(S_3) + \cdots.$$

The extension of μ proceeds as follows, from closed intervals to open sets, and it leads to an *approximation property* that will be the key to further extensions of μ.

1. From the definition $\mu([a, b]) = b - a$ it follows that

$$\mu(\{a\}) = \mu([a, a]) = a - a = 0.$$

   Thus, the measure of any singleton set is 0.
2. It follows, assuming subtractivity for intervals, that

$$\mu([a, b)) = \mu([a, b] - \{b\}) = \mu([a, b]) - \mu(\{a\}) = b - a = 0 = b - a,$$

   and similarly $\mu((a, b]) = \mu((a, b)) = b - a$. Thus, open, closed, and half-open intervals with the same endpoints have the same measure.
3. Assuming additivity for intervals, any finite disjoint union of intervals, with endpoints a_i, b_i such that

$$a_1 \leq b_1 \leq a_2 \leq b_2 \leq \cdots \leq a_n \leq b_n,$$

   has measure $(b_1 - a_1) + (b_2 - a_2) + \cdots + (b_n - a_n) \leq 1$.
4. An open set U is the union of countably many intervals (namely, intervals with rational endpoints), hence of countably many disjoint intervals by merging any

that overlap. If the nth interval has endpoints a_n, b_n then, assuming countable additivity for intervals,

$$\mu(U) = (b_1 - a_1) + (b_2 - a_2) + (b_3 - a_3) + \cdots$$
$$= \lim_{n \to \infty} [(b_1 - a_1) + (b_2 - a_2) + \cdots + (b_n - a_n)].$$

Since $(b_1 - a_1) + (b_2 - a_2) + \cdots + (b_n - a_n) \leq 1$ by the previous example, $\mu(U)$ exists as the limit of a bounded increasing sequence. (This example shows that the completeness of $\mathbb{R}$ is crucial to the theory of measure.)

5. It also follows, since $\lim_{n \to \infty} [(b_1 - a_1) + (b_2 - a_2) + \cdots + (b_n - a_n)]$ exists, that $(b_1 - a_1) + (b_2 - a_2) + \cdots + (b_n - a_n)$ is arbitrarily close to $\mu(U)$ for n sufficiently large. That is, *any open set can be approximated within measure ε by a finite union of intervals.*

Exercises

9.1.1 Prove by induction that Borel sets $S \subseteq [0, 1]$ differ from Borel subsets of $\mathcal{N}$ only by the presence of rational numbers.

9.1.2 Show that any closed set F is contained in a finite union F' of intervals that approximates F within measure ε.

9.2.3 Illustrate this result in the case of the Cantor set.

9.2 Approximation and Measure

The approximation property of open sets revealed in the last example—that any open set differs from a finite union of intervals by a union of intervals of total length $\leq \varepsilon$—is generalized in the following:

Definition. A set $S \subseteq [0, 1]$ is said to be *approximated within measure ε by a finite union F of intervals* if $(S - F) \cup (F - S)$ can be covered by intervals of total length $\leq \varepsilon$.

Thus, example 5 in the previous section shows that any open subset U of $[0, 1]$ can be approximated within measure ε by a finite union of intervals.

We now show that, for any $\varepsilon \geq 0$, each Borel set $S \subseteq [0, 1]$ can be approximated within measure ε by a finite union of intervals. The argument involves summing series like the series $\frac{\varepsilon}{2} + \frac{\varepsilon}{4} + \frac{\varepsilon}{8} + \cdots$ used at the beginning of this chapter. As usual, we argue by induction, proving that the property extends to Σ_α from the Σ_β with $\beta < \alpha$ via the operations of complementation and countable union. However, the ordinals are only incidental to the proof, so we will call this argument "induction on the construction of Borel sets."

In the proof that follows we take the word "interval" to mean any kind of interval: open, closed, or half-open. As we have seen, this makes no difference as far as measure is concerned. And it has the advantage that complement, union, and intersection of finite unions of intervals are again finite unions of intervals.

Borel approximation theorem. *For any $\varepsilon > 0$, each Borel set $S \subseteq [0, 1]$ can be approximated within measure ε by a finite union of intervals.*

Proof. We argue by induction on the construction of Borel sets, from open sets via complements and countable unions. The base step, where S is an open set, has already been done in example 5 of the previous section.

For the induction step, first suppose that S is the complement of an approximable Borel set $[0, 1] - S$. That is, $[0, 1] - S$ can be approximated within measure ε by a finite union F of intervals. It follows immediately (thanks to the symmetry of the definition) that S is approximated within measure ε by $[0, 1] - F$, which is also a finite union of intervals.

Finally, suppose that $S = S_1 \cup S_2 \cup S_3 \cup \cdots$, where each S_i is an approximable Borel set. In particular, we have finite unions of intervals

$$F_1 = \bigcup_k I_{1k}, \text{ which approximates } S_1 \text{ within measure } \varepsilon/4,$$

$$F_2 = \bigcup_k I_{2k}, \text{ which approximates } S_2 \text{ within measure } \varepsilon/8,$$

$$F_3 = \bigcup_k I_{3k}, \text{ which approximates } S_3 \text{ within measure } \varepsilon/16,$$

and so on. It follows that the union of all the intervals,

$$F = \bigcup_{j,k} I_{jk},$$

approximates S within measure $\varepsilon/2$, because

$$\frac{\varepsilon}{4} + \frac{\varepsilon}{8} + \frac{\varepsilon}{16} + \cdots = \frac{\varepsilon}{2}.$$

Of course, $F = \bigcup_{j,k} I_{jk}$ may be an infinite union of intervals. But some finite subset of these intervals has a union F' that approximates F within measure $\varepsilon/2$ (by the argument used for open sets in the previous section), so F' approximates S within measure ε.

This completes the induction.   □

Exercises

9.2.1 Verify that complement, union, and intersection of finite unions of intervals are again finite unions of intervals.

The set $(S - F) \cup (F - S)$ is called the *symmetric difference* of S and F, sometimes written $S \Delta F$. The proof above takes it for granted that if each $S_j \Delta F_j$ can be covered by intervals of total length at most $\varepsilon/2^{j+1}$ then

$$\bigcup_j S_j \Delta \bigcup_j F_j \text{ can be covered by intervals of total length at most } \sum_j \frac{\varepsilon}{2^{j+1}} = \frac{\varepsilon}{2}.$$

A more nitpicking proof may include the following details.

9.2.2 Observing that $\left(\bigcup_j S_j\right) - F = \bigcup_j(S_j - F)$ and $S_j - F \subseteq S_j - F_j$, deduce that $\bigcup_j S_j - \bigcup_j F_j \subseteq \bigcup_j(S_j - F_j)$.

9.2.3 Observing similarly that $\bigcup_j F_j - \bigcup_j S_j \subseteq \bigcup_j(F_j - S_j)$, deduce that $\bigcup_j S_j \Delta \bigcup_j F_j \subseteq \bigcup_j(S_j \Delta F_j)$.

9.3 Lebesgue Measure

The Borel approximation theorem prompts us to define measurability and measure of sets $S \subseteq [0, 1]$ as follows.

Definition. 1. A set $S \subseteq [0, 1]$ is *Lebesgue measurable* if, for any $\varepsilon > 0$, there is a finite union F of intervals that approximates S within measure ε.

2. The *Lebesgue measure* $\mu(S)$ of a Lebesgue measurable set S equals $\lim_{n\to\infty} \mu(F_n)$, where F_n is a finite union of intervals that approximates S within measure $1/n$.

Thus, Lebesgue measure extends the measure μ on intervals to all sets that can be approximated arbitrarily closely by finite unions of intervals. This definition of measure resembles the Greek "method of exhaustion" for measuring lengths and areas of curved figures—by approximating them with known figures, such as polygons.

It follows immediately from this definition and the Borel approximation theorem that *all Borel subsets of* $[0, 1]$ *are Lebesgue measurable*.

However, the Borel sets are far from being all subsets of $[0,1]$. We saw in Sect. 8.6 that there are only $2^{\aleph_0}$ Borel sets; that is, as many as there are points in $[0,1]$. And we know from Sect. 3.8 that $[0,1]$ has more subsets than elements. To find more measurable sets we take subsets of the Cantor set C, which is equinumerous with $[0,1]$ and of measure 0, by Sect. 3.7. It follows that C has as many subsets as $[0,1]$, and all of them are measurable, as subsets of a measure 0 set.

Thus, there are as many measurable sets as there are subsets of $[0,1]$. To decide whether *nonmeasurable* sets exist we therefore need to know more about Lebesgue measure than just the number of measurable sets. The properties of Lebesgue measure with the most bearing on this question are countable additivity, mentioned in Sect. 9.1, and *translation invariance*. The latter says that any measurable set S has the same measure as $S + r = \{x + r : x \in S\}$ (the "translate of S through distance r"). We now show why these properties hold for all measurable sets.

Properties of Lebesgue measure. *Lebesgue measure μ on $[0,1]$ is countably additive and translation invariant. That is:*

1. If $S_1, S_2, S_3, \ldots$ are disjoint measurable sets then

$$\mu(S_1 \cup S_2 \cup S_3 \cup \cdots) = \mu(S_1) + \mu(S_2) + \mu(S_3) + \cdots.$$

2. *If S is Lebesgue measurable and $S + r = \{x + r : x \in S\}$ then $\mu(S + r) = \mu(S)$.*

Proof. 1. As we showed in the proof of the approximation theorem, by approximating each S_n within $\varepsilon/2^{n+1}$ by a finite union of intervals I_{nk} we can approximate $S_1 \cup S_2 \cup S_3 \cup \cdots$ within ε by a finite union of intervals I_{jk}. Thus, the measure $\mu(S_1 \cup S_2 \cup S_3 \cup \cdots)$ can differ from the sum of measures $\mu(S_1) + \mu(S_2) + \mu(S_3) + \cdots$ by at most 2ε. Letting $\varepsilon \to 0$ we get

$$\mu(S_1 \cup S_2 \cup S_3 \cup \cdots) = \mu(S_1) + \mu(S_2) + \mu(S_3) + \cdots.$$

2. It is immediate from the definition $\mu([a,b]) = b - a$ that $\mu([a+r, b+r]) = b - a$. Hence, if $I_1, \ldots, I_k$ are intervals whose union approximates S within measure ε, then $I_1 + r, \ldots, I_k + r$ are intervals whose union approximates $S + r$ within ε. Consequently, $\mu(S + r) = \mu(S)$. $\qquad\qquad\square$

We have defined measurability only for subsets of $[0,1]$ for the sake of convenience: it ensures that measurability is the same as "having finite measure." We can also define measurability for subsets of $\mathbb{R}$; for example, by saying that S is measurable if and only if each set $S \cap [n, n+1]$ is of finite measure.

Exercises

Without loss of generality we can take all the intervals whose finite unions approximate measurable sets S to be closed. Thus, the finite union F_n that approximates S within measure $1/n$ is a closed set.

9.3.1 If we define $\lim_{n \to \infty} F_n$ to be $\{x : x \in F_n$ for all sufficiently large $n\}$, explain why $\lim_{n \to \infty} F_n$ differs from S by a set of measure 0.

9.3.2 Show that $\lim_{n \to \infty} F_n = \bigcup_m \bigcap_{n > m} F_n$ Hence, show that $\lim_{n \to \infty} F_n$ is a Σ_2 set.

9.3.3 Deduce that any measurable set differs by measure 0 from a Σ_2 set.

Other closure properties of the measurable sets may be established as countable union was.

9.3.4 Show closure under complement and that $\mu([0,1] - S) = 1 - \mu(S)$. Deduce closure under countable intersection.

Approximation by finite unions of intervals implies the following "0–1 law."

9.3.5 Suppose that a set $S \subseteq [0,1]$ has *uniform density* d in the sense that $\mu(S \cap I)/\mu(I) = d$ for each interval I. Show that $d = 0$ or $d = 1$.

9.4 Functions Continuous Almost Everywhere

One of the most important insights afforded by Lebesgue measure is that it is more important for properties to hold "almost everywhere" (i.e., everywhere but for a set of zero Lebesgue measure) than everywhere. For example, we have the theorem that any continuous function on a closed interval is Riemann integrable (Sect. 4.8). However, this does not completely characterize Riemann integrability because certain *dis*continuous functions are also Riemann integrable. Some of them look extremely discontinuous from a nineteenth-century point of view; for example, the *Thomae function* on [0,1]:

$$t(x) = \begin{cases} \frac{1}{n} & \text{if } x = \frac{m}{n} \text{ for integers } m, n \\ 0 & \text{if } x \text{ is irrational.} \end{cases}$$

As we saw in the exercises to Sect. 4.2, $t(x)$ is discontinuous at every rational point, so it has a dense set of discontinuities. However, $t(x)$ is continuous at all irrational points, which we now realize is *almost everywhere*. And, as we also saw in the exercises to Sect. 4.8, $t(x)$ is Riemann integrable.

This is no accident. *The bounded Riemann integrable functions on $[a, b]$ are precisely those that are continuous almost everywhere.* Thus, by relaxing "continuous" to "continuous almost everywhere" (together with the more trivial boundedness condition) we can exactly capture the concept of Riemann integrability.

To prepare for a proof of this characterization of Riemann integrability, in this section we explore the properties of almost everywhere continuous functions. This involves a loosening of the concept of continuity called α-continuity.

Definition. The function f is α-*continuous* at $x = c$ if there is a $\delta > 0$ such that

$$y, z \in (c - \delta, c + \delta) \Rightarrow |f(y) - f(z)| < \alpha.$$

Thus, there is a neighborhood of $x = c$ in which $f(x)$ varies by less than α. A few results follow immediately from this definition.

1. If f is continuous at $x = c$, then for every natural number n, there is a neighborhood of $x = c$ in which $f(x)$ varies by less than $1/n$. In other words, *for every n, f is $1/n$-continuous at $x = c$.*
2. Therefore, if f is *not* continuous at $x = c$, then for some n, f is not $1/n$-continuous at $x = c$.
3. So, if we let

$$D = \{x \in \mathbb{R} : f \text{ is not continuous at } x\},$$

$$D_\alpha = \{x \in \mathbb{R} : f \text{ is not } \alpha\text{-continuous at } x\},$$

then

$$D = \bigcup_{n=1}^{\infty} D_{1/n}.$$

4. Also, each set D_α is closed, because it contains all its limit points. Namely, if $c_1, c_2, c_3, \ldots \in D_\alpha$, then $f(x)$ varies by at least α in each neighborhood of each c_i. Consequently, if c is the limit of the c_i, then each neighborhood of c contains points c_i, and hence neighborhoods in which $f(x)$ varies by at least α. Thus, $c \in D_\alpha$ also.

Note that the set D of points of discontinuity may not be closed. The Thomae function, for example, is discontinuous at precisely the rational points. The sets D_α are easier to work with in this respect, and in the next section we will take advantage of the fact that D is a countable union of the closed sets $D_{1/n}$.

9.4.1 Uniform α-Continuity

Another ingredient we will need in the next section is an analogue of the theorem from Sect. 4.7—that continuity on a compact set implies uniform continuity. The analogue is that α-continuity on a compact set K implies uniform α-continuity on K, defined as follows.

Definition. A function f is *uniformly α-continuous* on the set S if there is a δ such that, for any $y, z \in S$,

$$|y - z| < \delta \Rightarrow |f(y) - f(z)| < \alpha.$$

The theorem then reads:

Compactness and uniform α-continuity. *An α-continuous function on a compact set is uniformly α-continuous.*   □

The proof is completely analogous to the proof for ordinary continuity.

Exercises

9.4.1 Find $D_{1/n}$ for the Thomae function $t(x)$.
9.4.2 Hence show that each $D_{1/n}$ is finite for $t(x)$.
9.4.3 Use $D = \bigcup_{n=1}^{\infty} D_{1/n}$ to give a new proof that the rationals in $[0,1]$ form a countable set.

An important class of functions that are continuous almost everywhere are the monotonic functions.

9.4.4 If $M(x)$ is monotonic, prove that $D_{1/n}$ has measure 0, hence that D has measure 0.

9.5 Riemann Integrable Functions

Now we are ready to prove Lebesgue's theorem that *a bounded function f on $[a, b]$ is Riemann integrable $\Leftrightarrow f$ is continuous almost everywhere*. We use the concept of α-continuity and the discontinuity sets D_α and D from the previous section. Thus, to say that f is continuous almost everywhere is to say that $\mu(D) = 0$ and, since $D = \bigcup_{n=1}^{\infty} D_{1/n}$, the latter condition is equivalent to $\mu(D_{1/n}) = 0$ for all n.

Without loss of generality we can replace $[a, b]$ by $[0, 1]$, because translating and rescaling the domain of a function does not affect its Riemann integrability. We also split the theorem into its two directions. The harder direction is:

Almost continuity implies Riemann integrability. *If f is bounded on $[0, 1]$ and $\mu(D) = 0$, then f is Riemann integrable.*

Proof. Suppose we have the bound $M \geq |f(x)|$ for f on $[0,1]$. We set

$$\alpha = \varepsilon/2$$

and first show that there are disjoint open intervals $I_1, I_2, \ldots, I_k$ of total length less than $\varepsilon/4M$ such that

$$D_\alpha \subseteq I_1 \cup I_2 \cup \cdots \cup I_k.$$

This is because D_α is compact and of measure 0. Since D_α is of measure 0, we can cover it by an open set (i.e., a union of open intervals I) of arbitrarily small measure, namely, the finite union F of intervals that approximate D_α within measure ε, plus the union of intervals of total length $\leq \varepsilon$ that cover $D_\alpha - F$. And since D_α is compact, finitely many of the intervals I also cover D_α, and the union of these is a *disjoint* union of certain open intervals $I_1, I_2, \ldots, I_k$. By choosing the open set covering D_α to have measure less than $\varepsilon/4M$ we then have

$$\mu(I_1 \cup I_2 \cup \cdots \cup I_k) < \varepsilon/4M.$$

Now let $K = [0, 1] - (I_1 \cup I_2 \cup \cdots \cup I_k)$. Since all points of K are outside the set D_α where f is not α-continuous, f is α-continuous on K. Also, K is closed (being the complement of the open set $I_1 \cup I_2 \cup \cdots \cup I_k$) and bounded, hence compact, so f is in fact *uniformly* α-continuous on K, by the theorem in the previous section.

Recall that uniform α-continuity means that there is a $\delta > 0$ such that, for any $y, z \in K$

$$|y - z| < \delta \Rightarrow |f(y) - f(z)| < \alpha = \varepsilon/2.$$

Thus, if we divide K into finitely many subintervals of length less than δ, the difference between the lub and glb of $f(x)$ in each is less than $\varepsilon/2$. Consequently, the difference between the upper and lower Riemann sums for f on K is also less than ε (since the length of K is at most 1).

Finally, consider upper and lower Riemann sums on $I_1 \cup I_2 \cup \cdots \cup I_k$. Since this set has length at most $\varepsilon/4M$ and $|f(x)| \leq M$, the difference between lub and glb of f on $I_1 \cup I_2 \cup \cdots \cup I_k$ is at most $2M$, hence the difference between upper and lower Riemann sums is at most

$$2M \cdot \varepsilon/4M = \varepsilon/2.$$

Adding the contributions from K and $I_1 \cup I_2 \cup \cdots \cup I_k$, we find that the difference between upper and lower Riemann sums for f over $[0,1]$ is at most ε. Since ε is arbitrary, this means f is Riemann integrable. $\square$

The easier direction is:

Riemann integrability implies almost continuity. *If f is Riemann integrable on $[0, 1]$, then the set D of discontinuities of f has measure 0.*

Proof. Since $D = \bigcup_{n=1}^{\infty} D_{1/n}$, it suffices to show that each D_α has measure 0. Given α and any $\varepsilon > 0$ we choose a partition of $[0,1]$ for which the difference between upper and lower Riemann sums for f is less than $\alpha\varepsilon$.

It follows that the intervals of the partition on which f varies by α or more have total length less than ε. These intervals cover the set D_α, so D_α is contained in a finite union of intervals with arbitrarily small total length. Thus, D_α has measure 0, as required. $\square$

Exercises

If f is continuous almost everywhere on $[a, b]$ then the same is true on any subinterval of the form $[a, x]$, hence the Riemann integral

$$F(x) = \int_a^x f(t)\,dt$$

exists for any x in $[a, b]$. We now investigate the extent to which the fundamental theorem of calculus holds for the almost continuous functions f. (We showed that it holds at all points x for continuous f in Sect. 4.8.)

9.5.1 Show that $F'(x) = \lim_{h \to 0} \int_x^{x+h} f(t)\,dt$ if the limit exists.
9.5.2 Show that the limit equals $f(x)$ almost everywhere.

9.6   Vitali's Nonmeasurable Set

The first example of a nonmeasurable set was discovered by Vitali (1905). Its existence depends on a fairly strong form of AC—a well-ordering of $\mathbb{R}$—but, in fact, all examples of nonmeasurable sets depend on rather strong forms of AC. We give another interesting example in the next section.

The Vitali nonmeasurable set. *Assuming that there is a well-ordering of $\mathbb{R}$, there is a set $V \subset [0,1]$ that is not Lebesgue measurable.*

Proof. The set V is most naturally viewed as a subset of the circle (of circumference 1), rather than as a subset of [0,1]. For this reason, we imagine [0,1] turned into a circle C by joining 0 to 1, so each $x \in [0,1)$ is associated with the point on the circle C at angle $2\pi x$.

In particular, each rational $q \in [0,1)$ corresponds to the point at angle $2\pi q$. We now call points x_1 and x_2 *equivalent* if $x_1 - x_2$ is rational. This equivalence relation partitions the circle C into *equivalence classes*, where

$$x_1, x_2 \text{ belong to the same class } \Leftrightarrow x_1 - x_2 \text{ is rational.}$$

For example, the rational points in [0,1) form one such class. If $x_1 - x_2$ is *not* rational then x_1 and x_2 belong to distinct classes, with no common point x (otherwise $x_1 - x$ and $x_2 - x$ would both be rational, in which case $x_1 - x_2$ would be rational too).

It follows from AC for subsets of $\mathbb{R}$ (and hence also from a well-ordering of $\mathbb{R}$) that there is a set V that includes exactly one member from each equivalence class E. In particular, V includes exactly one rational point. Now, for each rational q we let

$$V + q = \{x + q : x \in V\}.$$

Here $x + q$ denotes addition "on the circle," or mod 1, so that $x + q$ corresponds to the point on the circle obtained from the point at angle $2\pi x$ by rotating through angle $2\pi q$. Thus, $V + q$ is simply the result of rotating the set V through angle $2\pi q$.

We now observe that the sets $V + q$, for rational $q \in [0,1)$, have the following properties.

1. *The sets $V + q_1$ and $V + q_2$ are disjoint if $q_1 \neq q_2$.*

   If $x \in V + q_1$ and $x \in V + q_2$ then $x - q_1, x - q_2 \in V$, so $x - q_1$ and $x - q_2$ are either identical or inequivalent, because distinct members of V belong to distinct equivalence classes. It follows that q_1 and q_2 are identical or inequivalent. But q_1 and q_2 cannot be inequivalent, since their difference is rational, so we must have $q_1 = q_2$.

2. *The union of the sets $V + q$, over rational $q \in [0,1)$, is the whole circle C.*

   Any $x \in [0,1)$ is equivalent to some $x' \in V$, since V includes a member of each equivalence class. But then $x = x' + q$ for some rational q, in which case $x \in V + q$.

3. *If V is Lebesgue measurable, then so is each $V + q$, and $\mu(V + q) = \mu(V)$.*

   Because $V + q$ is a translate of V, and μ is translation invariant.

Now suppose that V *is* Lebesgue measurable, in which case we have either $\mu(V) = 0$ or $\mu(V) = \varepsilon > 0$.

If $\mu(V) = 0$, then the union of the sets $V + q$, for rational q, is a countable union of measure 0 sets by property 3, and hence also of measure 0. This is impossible, by property 2, since C has measure 1. If $\mu(V) = \varepsilon > 0$ then C is a countable *disjoint* union (by property 1) of sets of measure ε, so C has infinite measure, which is also impossible.

Thus, the necessary conclusion is that V is not Lebesgue measurable.   □

Exercises

A somewhat similar nonmeasurable set, again in the circle of circumference 1, may be obtained from an *irrational rotation* in place of the rational rotations in Vitali's example.

We let s be an irrational number, so adding s mod 1 amounts to rotation through $2\pi s$. Now prove the following.

9.6.1 For any point x, the points in

$$\text{orbit of } x = \{x, x \pm s, x \pm 2s, \ldots\}$$

are all different.

9.6.2 For any points x and y, the orbits of x and y are disjoint or identical.

9.6.3 If X (obtained by AC) includes exactly one point from each orbit, show that

$$X, \quad X \pm s, \quad X \pm 2s, \ldots$$

are disjoint sets that fill the circle.

9.6.4 Conclude, as in Vitali's example, that X is nonmeasurable.

9.7 Ultrafilters and Nonmeasurable Sets

A new kind of nonmeasurable set was introduced by Sierpiński (1938), based on the existence of nonprincipal ultrafilters, which had been proved by Tarski (1930). Sierpinski used an ultrafilter to construct a set $W \subset [0, 1]$ with the following two properties.

1. *For each $x \in [0, 1]$, $x \in W \Leftrightarrow 1 - x \notin W$.*

    Thus, in some sense, W includes half the points in $[0, 1]$, so we should have $\mu(W) = \frac{1}{2}$, if W is measurable at all.

2. *If $[0, 1]$ is divided into 2^m equal intervals I_k, then each $W \cap I_k$ is a translate of $W \cap I_1$, so $\mu(W \cap I_k) = \mu(W \cap I_1)$.*

    Thus, we should also have $\mu(W \cap I_k) = \frac{1}{2}\mu(I_k)$. This conflicts with the definition of measurable sets in Sect. 9.3, according to which they can be approximated within measure ε by finite unions of intervals.

The proof associates each set $X \in U$ with a number $x \in [0, 1]$, namely,

$$x = \sum_{n \in X} 2^{-n}.$$

However, in a countable number of cases, two sets X are associated with the same x. For example,

$$\frac{1}{2} = 2^{-1} = 2^{-2} + 2^{-3} + 2^{-4} + \cdots,$$

so the sets $\{1\}$ and $\{2, 3, 4, \ldots\}$ both give $x = \frac{1}{2}$. This situation occurs only for numbers x expressible as a finite sum of powers 2^{-n} (equivalently, numbers with a finite binary expansion). Since there are only countably many such x, the set of them has measure 0 and may be ignored. We do so in the proof below, where $[0, 1]^*$ denotes the interval $[0,1]$ minus the points with finite binary expansions.

The omitted points x may also be described as those with binary expansions that terminate in an infinite sequence of 1s. Thus, they correspond precisely to the cofinite sets X, which form the filter we extend to get U. The points $x \in [0, 1]^*$ that we consider are therefore those corresponding to the X in U that are *added* to the cofinite filter to make an ultrafilter.

Ultrafilter-based nonmeasurable set. *If U is an ultrafilter over $\mathbb{N}$ extending the cofinite ultrafilter, and if $x = \sum_{n \in X} 2^{-n}$, then the $x \in [0, 1]^*$ for $X \in U$ comprise a nonmeasurable set W.*

Proof. Since $[0, 1]^*$ omits all numbers that are finite sums of terms 2^{-n}, each $x \in W$ includes infinitely many terms. Each $x \in W$ also omits infinitely many terms 2^{-n}, since any x that omits only finitely many terms 2^{-n} can be rewritten as a finite sum of such terms.

Now $\sum_{n \in \mathbb{N}} 2^{-n} = 1$, so it follows for each $x \in W$ that

$$1 - x = 1 - \sum_{n \in X} 2^{-n} = \sum_{n \notin X} 2^{-n} \notin W,$$

since U is an ultrafilter and hence $X \in U \Leftrightarrow \mathbb{N} - X \notin U$. Thus,

$$x \in W \Leftrightarrow 1 - x \in [0, 1]^* - W,$$

which means that $[0, 1]^* - W$ is the reflection of W in $x = \frac{1}{2}$. So both sets have the same measure, if they are measurable at all.

We now prove the second property that follows from the measurability of W: *for any subdivision of $[0, 1]$ into 2^m equal subintervals I_k, $W \cap I_k$ has the same measure for each k.* In fact, we show that each $W \cap I_k$ is a translate of $W \cap I_1$. This follows by induction when we prove that *each $W \cap I_{k+1}$ is the translate of $W \cap I_k$ through distance 2^{-m}*, which amounts to proving

$$x \in W \Leftrightarrow x \pm 2^{-m} \in W. \tag{*}$$

To prove the latter claim, we go back to the definition of U, in order to show that

$$X \in U \Leftrightarrow X' \in U \quad \text{for any sets } X, X' \text{ that differ by a finite set.}$$

Certainly

$$X \in U \Rightarrow X \cup F \in U \quad \text{for any finite set } F,$$

because any filter is closed under supersets. But also

$$X \in U \Rightarrow X - F \in U \quad \text{for any finite set } F,$$

because

$$X - F = X \cap (\mathbb{N} - F) = X \cap (\text{cofinite set}) \in U,$$

since U includes all cofinite sets and is closed under intersections.

The consequent property of the numbers x is that

$$x \in W \Leftrightarrow x' \in W$$

for any x, x' whose expressions as $\sum 2^{-n}$ differ in finitely many terms. Thus, to prove (*) it suffices to prove that $x \pm 2^{-m}$ differs from x in only finitely many terms. This is easy to check (see exercises).

Now we are ready to decide whether W is measurable. If it is, we know that $\mu(W) = \frac{1}{2}$. In that case, since $\mu(W \cap I_k)$ is the same for each k, we have $\mu(W \cap I_k) = \frac{1}{2}\mu(I_k)$. This contradicts the definition of Sect. 9.3, that any measurable set can be approximated by a finite union of intervals. (In more detail: if W is measurable, approximate it within $\varepsilon/2$ by a finite union of intervals I. Then, by choosing m sufficiently large, approximate the union of the intervals I within $\varepsilon/2$ by a suitable union of intervals among the 2^m intervals I_k. This contradicts the second property of W, according to which only half the measure of the intervals I_k belongs to W.) $\square$

Exercises

Given $x = 2^{-n_1} + 2^{-n_2} + 2^{-n_3} + \cdots$ with $n_1 < n_2 < n_3 < \cdots$, we wish to show that $x' = x \pm 2^{-m}$ is a sum of distinct powers 2^{-n}, differing from the sum for x in only finitely many terms. This is obviously true for $x' = x + 2^{-m}$ when x does not include the term 2^{-m}, and for $x' = x - 2^{-m}$ when x does include the term 2^{-m}. We now consider the remaining cases.

9.8.1 Suppose that $x = 2^{-n_1} + 2^{-n_2} + 2^{-n_3} + \cdots$, with $n_1 < n_2 < n_3 < \cdots$, includes the term 2^{-m}. Explain why $x' = x + 2^{-m}$ can be written as a sum of distinct powers 2^{-n} that differs from the sum for x only in terms 2^{-n} with $n \leq m$.

9.8.2 Explain why $2^{-1} - 2^{-4} = 2^{-2} + 2^{-3} + 2^{-4}$.

9.8.3 If $n < m$, show that

$$2^{-n} - 2^{-m} = 2^{-(n+1)} + 2^{-(n+2)} + \cdots + 2^{-m}$$

9.8.4 Suppose that $x = 2^{-n_1} + 2^{-n_2} + 2^{-n_3} + \cdots$ with $n_1 < n_2 < n_3 < \cdots$ and that x does not include the term 2^{-m}. Show, using Exercise 9.8.3 or otherwise, that $x' = x - 2^{-m}$ can be written as a sum of distinct powers 2^{-n} that differs from the sum for x in only the terms with $n \leq m$.

The proof above—that W is not measurable—assumes "reflection invariance" of Lebesgue measure in claiming that W has the same measure as $[0, 1]^* - W$.

9.8.5 Formulate "reflection invariance" precisely, and explain why it holds for the Lebesgue measure.

9.8   Historical Remarks

The two founders of measure theory, as we know it today, were Borel and Lebesgue. As mentioned in Sect. 7.9, Borel (1898) saw that it is natural to expect countable additivity for any concept of measure, and that this implies the measurability of all Borel sets. He also emphasized the immediate consequence of countable additivity, that countable sets have measure 0; in other words, that all sets of positive measure are uncountable. Finally, he was aware of the potential *inconsistency* of the measure concept—in the sense that different constructions of the same set could give different values of its measure—and he saw that inconsistency is averted by the Heine–Borel theorem. An interval of positive measure, such as $[0,1]$, cannot be covered by intervals I_n of total measure ε, since the Heine–Borel theorem then gives a covering of $[0,1]$ by *finitely many* I_n of total measure at most ε, which is clearly impossible.

This is why Borel called the Heine–Borel theorem the "first fundamental theorem of measure theory." Borel (1950) also spoke of the "second fundamental theorem of measure theory," meaning the fact that every measurable set can be approximated within ε by a finite union of intervals. Lebesgue (1902) took this property as the definition of a measurable set in $[0,1]$, much as we did in Sect. 9.3, and used it to prove the fundamental properties of measure, such as countable additivity.

Lebesgue extended Borel's ideas on measure in two ways. First, he extended the concept of measure beyond the Borel sets by including all *subsets* of measure 0 sets among the measure 0 sets. As we now know, this extends measure to all subsets of $[0,1]$, except for sets called into being by AC. Second, Lebesgue applied the concept of measure to analysis, particularly to the theory of integration. Lebesgue's concept of integral, now called the *Lebesgue integral*, is an extension of the Riemann integral to a much larger class of functions. Just as Lebesgue measure covers all Borel sets plus sets that are "almost Borel," the Lebesgue integral covers all bounded Baire functions plus such functions that are "almost Baire." This includes the Dirichlet function, and many other functions that are not Riemann integrable. Moreover, the Lebesgue integral has various *limit properties* that fail for the Riemann integral. Among them are:

Monotone Convergence Theorem.   If $f_1 \le f_2 \le f_3 \le \cdots$ is a sequence of Lebesgue integrable functions that converge almost everywhere on $[a, b]$ to f, and if $\int_a^b f_n(x)\, dx < A$ for each n, then

$$\int_a^b f(x)\, dx = \lim_{n \to \infty} \int_a^b f_n(x)\, dx.$$

Dominated Convergence Theorem.   If $f_1, f_2, f_3, \ldots$ are Lebesgue integrable functions that converge almost everywhere on $[a, b]$ to f, and if there is a Lebesgue integrable g with $|f_n| \le g$ almost everywhere in $[a, b]$, then f is Lebesgue integrable and

$$\int_a^b f(x)\, dx = \lim_{n \to \infty} \int_a^b f_n(x)\, dx.$$

Both of these theorems fail for the Riemann integral if we take

$$f_n(x) = \begin{cases} 1 \text{ on the first } n \text{ rational points } x \\ 0 \text{ elsewhere,} \end{cases}$$

so $\lim_{n \to \infty} f_n$ is the Dirichlet function, which is not Riemann integrable, even though each Riemann integral $\int_a^b f_n(x)\, dx = 0$.

Lebesgue's concept of "almost everywhere" corrects many cases of bad behavior previously thought to be incorrigible. We have already seen the example of the Thomae function, which looks badly discontinuous, but is actually continuous almost everywhere. Other cases of badness shown to be "good almost everywhere" by Lebesgue concern the differentiability of continuous functions and the fundamental theorem of calculus:

- Any monotonic continuous function is differentiable almost everywhere.
- If f is integrable and $F(x) = \int_a^x f(t)\, dt$ then F is differentiable almost everywhere and $F'(x) = f(x)$ almost everywhere.

Lebesgue's results changed the face of analysis by greatly expanding the scope of the operations of integration and differentiation. They also drew attention to the underlying concept of measure, and raised the question of nonmeasurable sets. The example of Vitali (1905) was challenging in its simplicity, despite its reliance on AC, which Lebesgue did not accept. More challenging examples were to come.

Hausdorff (1914), p. 469, gave a decomposition of the sphere into three pieces A, B, and C such that:

- A is congruent to B (by a rotation of the sphere),
- A is congruent to C (by a rotation of the sphere),
- A is congruent to $B \cup C$ (by a rotation of the sphere).

Fig. 9.1 Giuseppe Vitali

It follows that subsets of the sphere cannot even be given a *finitely* additive measure μ, if measure is assumed to be rotation-invariant.[1] If they could, we should have nonzero numbers satisfying the contradictory equations

$$\mu(A) = \mu(B) = \mu(C) = \mu(B) + \mu(C).$$

Hausdorff's sets A, B, and C are determined with the help of AC. Also using AC, Banach and Tarski (1924) used Hausdorff's idea to devise the ultimate affront to common sense: a decomposition of the three-dimensional unit ball into a finite number of subsets, which can be rigidly moved to form *two* unit balls. Accounts of this "Banach–Tarski paradox" may be found in Wagon (1993) and Wapner (2005). Figure 9.2 shows Banach and Tarski around the year 1919, five years before the Banach–Tarski theorem.

As mentioned in Chap. 6, Gödel (1938) proved that AC is consistent with the ZF axioms, so the Banach–Tarski paradox is not a contradiction (unless the ZF axioms are themselves contradictory). Nevertheless, one might hope that there are other axioms with some of the benefits of AC without its counterintuitive consequences. The most deeply studied candidate so far is the axiom of determinacy, AD, which we introduced in Chap. 7.

There we mentioned that AD has some benefits for the theory of $\mathbb{R}$, because it implies that all subsets of $\mathbb{R}$ are Lebesgue measurable and that countable AC holds for sets of reals. What distinguishes AD from AC is its "higher consistency strength," meaning that we have to assume the existence of very large sets to prove the consistency of ZF+AD. Since AC is normally assumed in these consistency

---

[1]More generally, we would like measure to be invariant under any rigid motions, such as translations, reflections, and rotations. This is the case for Lebesgue measure in all dimensions.

Fig. 9.2 Stefan Banach and Alfred Tarski

proofs, one can measure large sets by their cardinal numbers, which means we assume the existence of *large cardinals*.

Recall from Sect. 7.9 that the Gödel (1938) theorem about AC is that ZF+AC is consistent provided only that ZF is consistent. This means that ZF+AC has the same consistency strength as ZF. A proposition of somewhat higher consistency strength is "all subsets of $\mathbb{R}$ are Lebesgue measurable." As mentioned in Sect. 6.8, to prove the consistency of this proposition one needs to assume the existence of an inaccessible cardinal—something not provable in ZF alone. Inaccessible cardinals are merely the smallest of the large sets whose existence cannot be proved in ZF. We have to assume the existence of much larger sets, called *Woodin cardinals*, to prove the consistency of ZF+AD (again, assuming that ZF is consistent). This result follows from the theorem of Woodin, mentioned in Sect. 7.9, that AD holds in $L(\mathbb{R})$.

Another way to deal with the paradoxical sets is to allow them to exist (by assuming AC), but to keep them as far as possible from sets we can define, such as the Borel sets and those obtained from them by projection and complementation: the *projective* sets. Woodin cardinals also solve this problem. If we assume the consistency of

$$ZF+AC+\text{"there are infinitely many Woodin cardinals,"}$$

then we get the consistency of

$$ZF+AC+\text{"all projective sets are determined."}$$

This was proved by Martin and Steel (1989). As mentioned in Sect. 7.9, Mycielski and Świerczkowski (1964) proved that all determined sets are measurable, so Martin and Steel's result keeps the nonmeasurable sets outside the projective sets, provided those infinitely many Woodin cardinals exist.

It is perhaps frustrating to be unable to answer basic questions about the real numbers without assuming the existence of astoundingly large sets. But it is surely inspiring to know that the human mind can find such a remarkable explanation for the gaps in our understanding of $\mathbb{R}$.

Chapter 10
Reflections

PREVIEW

In this chapter we revisit the fundamental questions raised in the first chapter. We review the answers obtained, and reflect on the insights and new questions to which they lead.

The fundamental questions were already implicit in ancient Greek mathematics, and the Greeks saw that their difficulties were entangled with the concept of infinity—which worried them. On the other hand, they also saw that infinity could be used to solve otherwise unapproachable problems, such as finding the area of a parabolic segment. But their qualms about infinity prevented them from using infinite processes systematically (i.e., from developing calculus).

We now realize that the difficulties of Greek mathematics are concentrated in the concept of a *real number*: a concept that has to meet the needs of both arithmetic (counting, adding, multiplying) and geometry (measuring quantities such as length and area, modeling continua such as lines and curves).

To reconcile these demands requires acceptance of infinity, and with it the general concepts of set, function, and limit—none of which were known to the Greeks. Set theory is a setting where questions about infinity, functions, and limits can be answered to a large extent. In particular, ZF set theory (plus certain axioms of choice) has given good answers to most of the fundamental questions. But it has also shown the questions to be more complicated than first thought. In particular, the problem of measuring sets of real numbers is entangled with questions about the entire universe of infinite sets.

J. Stillwell, *The Real Numbers: An Introduction to Set Theory and Analysis*,
Undergraduate Texts in Mathematics, DOI 10.1007/978-3-319-01577-4_10,
© Springer International Publishing Switzerland 2013

10.1   What Are Numbers?

The answer to this question turns out to have two parts:

1. The laws of arithmetic, which govern the behavior of addition, subtraction, multiplication, and division, arise from the *positive integers* $1, 2, 3, 4, 5, \ldots$, and the accompanying principle of *induction*. These laws extend quite easily to the *integers* $\ldots -2, -1, 0, 1, 2, 3, \ldots$ and the *rational numbers* m/n (where m and n are integers and $n \neq 0$).
2. Since irrational numbers exist—for example, $\sqrt{2}$ and π—we need to define irrational numbers and to extend the laws of arithmetic to them. A convenient way to do this is to define each positive *real number* as a *cut* in the set of positive rational numbers; that is, a partition of the set $\mathbb{Q}^+$ of rational numbers ≥ 0 into sets L and U, where each member of L is less than each member of U.

   Each rational number r is thereby represented by the cut $\langle L, U \rangle$, where

$$L = \{q \in \mathbb{Q}^+ : q \leq r\}, \quad U = \{q \in \mathbb{Q}^+ : q > r\},$$

or by the cut $\langle L', U' \rangle$, where

$$L' = \{q \in \mathbb{Q}^+ : q < r\}, \quad U = \{q \in \mathbb{Q}^+ : q \geq r\}.$$

A cut $\langle L, U \rangle$ in which L has no maximum member and U has no minimum member therefore represents an *irrational* number. For example, $\sqrt{2}$ is represented by the cut

$$L_{\sqrt{2}} = \{q \in \mathbb{Q}^+ : q^2 < 2\}, \quad U_{\sqrt{2}} = \{q \in \mathbb{Q}^+ : q^2 > 2\}.$$

Thus, the rational numbers are reinvented, and the irrational numbers are brought into being, as (pairs of) infinite sets of positive rational numbers. The laws of arithmetic are then *inherited*, as it were, from the rational numbers. It suffices to define the *sum of cuts* $\langle L, U \rangle$ and $\langle L', U' \rangle$ as the cut with lower set

$$L + L' = \{q + q' : q \in L \text{ and } q' \in L'\},$$

and the *product of cuts* is the cut with lower set

$$LL' = \{qq' : q \in L \text{ and } q' \in L'\}.$$

The laws of arithmetic for positive real numbers extend to zero and negative real numbers without difficulty. One can also prove properties of algebraic numbers, such as $(\sqrt{2})^2 = 2$ and $\sqrt{2}\sqrt{3} = \sqrt{6}$.

The step from the rational to the real numbers is accomplished by a single axiom of ZF set theory: the axiom of infinity. More precisely, the theory of rational numbers under addition and multiplication is essentially the same as ZF–Infinity, the theory of finite sets. As we explained in Sect. 6.6, the natural numbers can be taken to be the finite sets

$$0 = \{\}, \quad 1 = \{0\}, \quad 2 = \{0, 1\}, \quad \ldots .$$

The successor function is then $S(n) = n \cup \{n\}$, and $+$ and $\times$ can be defined by induction as explained in Sect. 2.2. This embeds the arithmetic of natural numbers in the theory of finite sets, and one can extend the theory to integers and rational numbers by using ordered pairs, as was also explained in Sect. 6.6. Conversely, all finite sets can be encoded as natural numbers, and operations on finite sets, such as pairing and union, can be simulated by arithmetic operations.

The axiom of infinity takes us from $\mathbb{Q}$ to $\mathbb{R}$ and of course much more. ZF set theory includes objects far beyond individual real numbers, or the set $\mathbb{R}$, or the subsets of $\mathbb{R}$, or the real functions that form the subject matter of analysis. Nevertheless, in passing from ZF–Infinity to ZF we do not necessarily overstep the bounds of analysis. If anything, we need *more* axioms of set theory, not less, to answer questions about $\mathbb{R}$. As we have seen, some form of choice axiom is needed to establish the equivalence of ordinary continuity and sequential continuity. And some questions about measure and determinacy of subsets of $\mathbb{R}$ cannot be settled without appealing to large cardinal axioms, that is, assumptions about the size of sets that are able to exist.

10.2 What Is the Line?

While Dedekind's concept of cut provides an immediate extension of the laws of arithmetic from rational to irrational numbers, that was not his main reason for introducing it. What he really wanted to do was provide a numerical model of the line, and particularly its "continuity," or what we now call its *completeness*. The rational numbers do not provide a convincing model of the line because they have *gaps* at places such as $\sqrt{2}$. Gaps in the line stymie any attempt to prove basic theorems about continuous functions, such as the intermediate value theorem.

For example, the function $f(x) = x^2 - 2$ continuously passes from a negative value (-2) at $x = 0$ to a positive value $(+2)$ at $x = 2$, without taking the value 0 at any rational value of x. Thus, even a quadratic function on $\mathbb{Q}$ may fail to have the intermediate value property.

Dedekind's way of creating a line without gaps is almost absurdly simple: fill each gap in the rational numbers by *the gap itself*! And what better way to realize a gap in the rationals than as the pair $\langle L, U \rangle$ of sets of rationals, respectively, to the left, and to the right, of the gap? What made Dedekind's idea so revolutionary was its introduction of *sets* as bona fide mathematical objects. This was too radical for Dedekind's contemporaries, and perhaps even for Dedekind himself, because he claimed the right to *create* a new number corresponding to each pair $\langle L, U \rangle$. Today however, we are used to defining mathematical objects as sets, so a pair $\langle L, U \rangle$ is as good a mathematical object as any other.

Realization of gaps by sets of rational numbers may be the simplest way to construct a line without gaps, but the set $\mathbb{R}$ of real numbers is not as simple as the set $\mathbb{Q}$ of rational numbers. As we saw in Chap. 3, $\mathbb{Q}$ is countable, like the set of natural numbers, but $\mathbb{R}$ is not. This makes a sharp distinction between rational and real numbers, which was not clear when we casually made the step from rational numbers to *sets of* rational numbers. The uncountability of $\mathbb{R}$ exposes the enormity of this step, and more generally of the step from any set S to its power set $\mathcal{P}(S) = \{$all subsets of $S\}$. As we saw in Sect. 3.8, this step always leads to a set of higher cardinality, so the power set axiom of ZF—which allows us to view sets as *objects*, and hence as members of another set—is not to be taken lightly.

10.3   What Is Geometry?

Since ancient times, geometry has been based on the idea of a continuous space (originally the line, the plane, or three-dimensional space) with a distance function. Distance is the fundamental geometric quantity, since it determines all other geometric quantities, such as angle, area, and volume. We now know that the idea of a continuous line can be modeled by $\mathbb{R}$, the set of real numbers, and $\mathbb{R}$ also gives a distance between any two points x and y in $\mathbb{R}$; namely, $|y - x|$.

The plane can then be modeled by the cartesian product of the line with itself,

$$\mathbb{R}^2 = \{\langle x_1, x_2 \rangle : x_1, x_2 \in \mathbb{R}\},$$

which goes back to the idea (as you can see from the word "cartesian") of Descartes (1637) of using coordinates x_1, x_2 to describe points in the plane. As we saw in Sect. 5.1, the Pythagorean theorem motivates the definition of distance between points $\langle x_1, x_2 \rangle$ and $\langle y_1, y_2 \rangle$, namely,

$$\sqrt{(y_1 - x_1)^2 + (y_2 - x_2)^2}.$$

More generally, we define the distance between points $\langle x_1, x_2, \ldots, x_n \rangle$ and $\langle y_1, y_2, \ldots, y_n \rangle$ in $\mathbb{R}^n$ to be

$$\sqrt{(y_1 - x_1)^2 + (y_2 - x_2)^2 + \cdots + (y_n - x_n)^2},$$

motivated by the idea of iterating the Pythagorean theorem (using a triangle with one side in $\mathbb{R}^{n-1}$ and a perpendicular side into $\mathbb{R}^n$). The generalization of distance to n dimensions is elegantly subsumed by the concept of *inner product* on $\mathbb{R}^n$:

$$\langle x_1, x_2, \ldots, x_n \rangle \cdot \langle y_1, y_2, \ldots, y_n \rangle = x_1 y_1 + x_2 y_2 + \cdots + x_n y_n.$$

We abbreviate the points $\langle x_1, x_2, \ldots, x_n \rangle$ and $\langle y_1, y_2, \ldots, y_n \rangle$ by x and y, the distance between x and y by $|y - x|$, and we further abbreviate the distance $|x - 0|$ between x and the origin 0 by $|x|$.

Then we notice that $|x|^2 = x \cdot x$ and, more generally,

$$|y - x|^2 = (y - x) \cdot (y - x).$$

Thus, distance is expressible in terms of the inner product. So too is angle, because it happens that

$$x \cdot y = |x||y| \cos \theta,$$

where θ is the angle between the lines from the origin to x and y. The inner product thereby gives the basic concepts of geometry on $\mathbb{R}^n$. In this geometry the Pythagorean theorem holds—because it is built into the definition of inner product—and also the other fundamentals of geometry laid down by Euclid. For this reason, $\mathbb{R}^n$ with its inner product is called *Euclidean space*.

It is noteworthy how much of this realization (no pun intended) of Euclid's vision builds on ideas already present in Euclid. The Pythagorean theorem is the main theorem of Euclid's *Elements*, Book I, and the struggle to incorporate irrationals into the line is the subject of Euclid's Book V. What Euclid lacked, as we have seen, was sufficient acceptance of infinity to obtain a complete *number* line $\mathbb{R}$, and sufficient acceptance of algebra to admit products of any number of lengths. [Incidentally, it was Grassmann (1847) who first proposed algebraic foundations for n-dimensional geometry. Thus, Grassmann was a pioneer in the foundations of both arithmetic and geometry.]

The linear algebra courses of today, which often present the definition of Euclidean space without comment, stand upon the shoulders of giants: Euclid, Descartes, Grassmann, Dedekind, Cantor, ... And this is just Euclidean geometry. Since the early nineteenth century, there have also been *non*-Euclidean geometries, typically based on $\mathbb{R}$ as well.

Historically, Euclidean geometry was first generalized by considering surfaces in $\mathbb{R}^3$, and measuring distance between two points on the surface in the natural way. For example, if one has the unit sphere $\mathbb{S}^2$ in $\mathbb{R}^3$ one wants to measure the distance between points P and Q on $\mathbb{S}^2$ by taking the plane through P, Q and the center O of $\mathbb{S}^2$, which meets $\mathbb{S}^2$ in a circle of radius 1, and measuring the length of the arc from P to Q on this circle.

More generally, on a smooth surface S in $\mathbb{R}^3$ one hopes that for any two points P and Q on S there will be a *geodesic* (curve of shortest length) connecting P and Q, the length of which can be found by calculus.

More generally still, one can dispense with the ambient space $\mathbb{R}^3$ or $\mathbb{R}^n$ altogether and simply define a length function $d(P, Q)$ on any space S to be a real-valued function with reasonable properties. The most general properties that are geometrically reasonable are the following: for all $P, Q \in S$

1. $d(P, Q) \geq 0$ (distance is positive),
2. $d(P, Q) = 0 \Leftrightarrow P = Q$ (points at zero distance are identical),
3. $d(P, Q) = d(Q, P)$ (symmetry), and
4. $d(P, R) \leq d(P, Q) + d(Q, R)$ (the triangle inequality).

These properties define what is called a *metric space*, which is a general setting for geometry based on a real-valued length function.

10.4 What Are Functions?

The simplest answer is based on the concept of a set, as we already saw in Sect. 1.5. However, we now know that the simplicity of this general definition has a cost. Real functions become as complicated as sets of real numbers, so their properties depend to some extent on which axioms of set theory we accept.

For example, with a well-ordering of $\mathbb{R}$ we have a nonmeasurable set $V \subseteq [0, 1]$ (the Vitali set), and the characteristic function of V is not Lebesgue integrable. If, on the other hand, we accept the axiom of determinacy (AD), then all subsets of [0,1] are Lebesgue measurable and all (bounded) functions on [0,1] are Lebesgue integrable.

There are also options between these two extremes, concerning the *complexity* possible for nonmeasurable functions. To explain them we recall the concepts of Baire function and Borel set from Chap. 8.

The Baire functions are those obtainable from continuous functions by the limit operation, and they include the characteristic functions of all sets in the Borel hierarchy. The Borel sets are all Lebesgue measurable, as we saw in Sect. 9.3, and this enables us to prove that all (bounded) Baire functions on [0,1] are Lebesgue integrable. Thus, nonintegrable functions have greater complexity than Baire functions. But we can go a little further.

We can also prove (in ZF+AC) that all *projections* of two-dimensional Borel sets are measurable. It turns out that this class of sets (called analytic sets) includes sets that are not Borel, so Lebesgue measurability extends beyond the Borel sets, to the analytic sets and their complements. These sets form the first level of what is called the *projective hierarchy*, whose higher levels arise by taking finitely many complements and further projections.

It follows from the measurability of analytic sets and their complements that the corresponding functions are Lebesgue integrable. However, in ZF+AC we can*not* prove measurability of all sets at higher levels of the projective hierarchy, so Lebesgue integrability of the corresponding functions is also not provable. Thus, the question of how complex a nonintegrable function must be is essentially the question of how complex a nonmeasurable set must be. We pursue this question further in Sect. 10.6.

Suppose, however, that one gives up this pursuit and decides to confine attention to the Baire functions. Even here we cannot avoid set theory issues. Just as the concept of natural number is naturally linked to induction over the finite ordinals, the Baire functions (and the related Borel sets) are naturally linked to the countable ordinals and transfinite induction. And in this domain we cannot avoid using countable AC; for example, to prove that a countable union of countable sets is countable.

10.5 What Is Continuity?

In ordinary speech, the word "continuous" means "unbroken" or "without gaps," as in "a continuous curve" or "continuous progress." In mathematics we now use the term "connected" for this property, and reserve the term "continuous" for functions—though the two concepts are certainly related. A continuous function f on $\mathbb{R}$ has a connected graph, for example, but this is due as much to the completeness of $\mathbb{R}$ as to the continuity of f.

Analysis suggests two definitions of continuity: the ε-δ definition and the definition via sequences ("sequential continuity"). The two are equivalent only by virtue of an axiom of choice, namely countable AC, which is needed to prove that sequential continuity at a point implies ε-δ continuity at a point. The latter result was one of the first to raise issues of set theory in analysis.

However, neither of these definitions was as consequential as the third definition, in terms of open sets. Hausdorff's discovery that a function f is continuous if and only if f^{-1} of any open set is open, became the foundation of the whole discipline of *topology*. One begins with the concept of a *topological space* S, which is a set together with a collection $\mathcal{T}$ of subsets U that are called *open*. The sets in $\mathcal{T}$ are open purely by virtue of the following closure properties:

1. The empty set and the whole space S are open.
2. If U and V are open, then $U \cap V$ is open.
3. If $\{U_i\}$ is a collection of open sets, then $\bigcup_i U_i$ is open.

In this setting, one can now talk about continuous functions, homeomorphisms, closed sets, compact sets, and so on, because all are definable in terms of open sets. $\mathcal{T}$ is called a *topology on* the set S.

The generality of the continuity concept means that problems about the existence of continuous maps can be quite difficult. An example is the problem of invariance of dimension (nonexistence of a continuous bijection $\mathbb{R}^m \to \mathbb{R}^n$ for $m \neq n$), which is nontrivial even for $m = 1$ and $n = 2$.

10.6 What Is Measure?

As already mentioned in Sect. 10.4, the concept of measure leads to questions about functions (such as integrability) which have answers that depend on which axioms of set theory we accept. Thus, the concept of measure has been decisive in exposing the role of set theory in analysis. Our original question (What is measure?) can be answered fairly simply: it is what you get by approximating a set by finite unions of intervals (as explained in Sect. 9.3).

The real question is: What is a measurable set? In other words, which sets *can* be approximated by finite unions of intervals? The answer depends on our axioms for set theory, and there are conflicting answers for AC versus AD.

With full AC, there are nonmeasurable sets, as we saw in Sects. 9.6 and 9.7. However, we cannot establish the complexity of these nonmeasurable sets without axioms that go beyond ZF+AC. The lowest possible complexity of nonmeasurable sets results from adding the axiom of constructibility due to Gödel (1939). As mentioned in Sect. 7.3, this axiom says that each set has a definition in a language that includes symbols for ordinals (in addition to the symbols in the usual language for set theory). It follows that the definitions of sets can be well-ordered, which leads to an explicitly defined well-ordering of $\mathbb{R}$. From this definition one can obtain definitions of nonmeasurable sets at the second level of the projective hierarchy. This is as low as we can go because sets at the first level can be proved measurable in ZF+AC, as we mentioned in Sect. 10.4.

It is not known how high one can push the nonmeasurable sets while still retaining AC. The strongest result so far is that all projective sets can be proved measurable if we add to ZF+AC an axiom stating the existence of very "large" sets. This was proved by Martin and Steel (1989). As mentioned in Sect. 9.8, they actually proved *determinacy* of projective sets from the assumption that these "large" sets exist, whence measurability follows from the theorem of Mycielski and Świerczkowski (1964) that determinate sets are measurable.

Thus, the theorem of Martin and Steel (1989) is also a theorem about the extent to which AD conflicts with AC: it says that AC is compatible with projective AD. In that respect it complements the much easier theorem proved in Sect. 7.7, which says that full AD is compatible with countable AC for sets of reals.

10.7 What Does Analysis Want from $\mathbb{R}$?

Analysis wants to apply the limit concept to numbers and functions, so it wants limits to exist among the real numbers wherever possible. That is, it wants $\mathbb{R}$ to be *complete*. As we have seen, this is equivalent to the geometrical demand of having "no gaps" in the line, so in this respect geometry and analysis make the same demands on $\mathbb{R}$. However, analysis has additional demands, such as having real numbers as the values of measures and integrals. As we saw in the previous section, this leads to the delicate question of deciding which sets are measurable. Deciding which functions are integrable is essentially the same question.

A related, though less delicate, role for the real numbers is to serve as values of the *distance function* in a *metric space*. The completeness of $\mathbb{R}$ is also important in a metric space S, for example, in defining the lengths of curves in S. If C is a curve in S given by a continuous function $f : [0, 1] \to S$ it is natural to approximate C by

a *polygon* with vertices $P_0 = f(0), P_1 = f(x_1), P_2 = f(x_2), \ldots, P_n = f(1)$ that lie
on C (with $0 < x_1 < x_2 < \cdots < 1$), and to find the length of C by finding a limiting
value of the polygon length

$$d(P_0, P_1) + d(P_1, P_2) + \cdots + d(P_{n-1}, P_n)$$

as the minimum side length tends to zero. If the limiting value exists, then the curve
C is said to be *rectifiable*.

As the latter example shows, analysis (and topology) also wants $\mathbb{R}$ to serve as
a model for continuous curves. That is, $\mathbb{R}$ (or $[0,1]$) is supposed to serve as the
domain of the continuous function that defines the curve. Due to the generality of
the concept of a continuous function, some "pathological" curves are admitted by
this definition, such as space-filling curves and curves with no tangents. The latter
curves also do not have finite length—if they have "length" at all, it is infinite.

The concept of *differentiability* is one way to restrict the class of curves to
more "natural" examples, since a differentiable curve has a tangent at each point
by definition. (Indeed, differentiability can be viewed as the property of a curve
approximating its tangent under indefinite magnification.) One also finds that
differentiable curves are rectifiable, and there is a natural formula for their length.
But here, too, the completeness of $\mathbb{R}$ is crucial. Length is the limit of a sequence of
real numbers, and one needs completeness to be sure that the limit exists.

10.8 Further Reading

The following, mostly modern, books are suggested for further exploration of the
ideas in this book. I believe that it will also help to dip into the classics, most
of which are available in English translation. Many translations are listed in the
bibliography, and a particularly good anthology of them is Ewald (1996).

10.8.1 Greek Mathematics

The works of Euclid and Archimedes contain the most important mathematics that
survives from ancient Greece, and they may be read in several different editions.
There are also some important works inspired by Euclid, such as the Hilbert (1899)
Foundations of Geometry, which analyzes the logic of Euclid's geometry and fills
its gaps.

Heath's 1925 edition of Euclid's *Elements*, reprinted as Euclid (1956), is
beginning to show its age, but it is still widely available and worth reading, if only
for Heath's rich and extensive commentary.

Two valuable supplements to Euclid are the books of Artmann (1999) and
Hartshorne (2000). Artmann, like Heath, is well versed in the history of Greek

mathematics, but with a more modern perspective. Hartshorne focuses on the transformation of Euclid's ideas by Hilbert, giving a rigorous, modern approach to geometry and its algebraic foundations.

As far as Archimedes is concerned, the most complete edition is still that of Heath (1897). Like Heath's Euclid, it is still worth reading, though it may soon be replaced as a result of recent scholarship. Reviel Netz is preparing a new three-volume edition, and one volume has appeared so far: Archimedes (2004).

10.8.2 The Number Concept

An excellent book devoted entirely to the number concept (and its higher-dimensional generalizations) is *Numbers* by Ebbinghaus et al. (1991). The first two chapters, in particular, give an expanded account of what was covered in Chap. 1 of this book.

After that, you may dare to read Landau (1951). Although entitled *Foundations of Analysis*, the book is really about foundations for the real and complex numbers (though I agree that this is an important part of the foundations of analysis).

10.8.3 Analysis

Understanding Analysis by Abbott (2001) is a pleasant undergraduate text that makes good use of the sequential continuity concept, though without discussing the related issues of set theory. Abbott also has a nice treatment of functions continuous almost everywhere and the Riemann integral, which I have drawn on in Chap. 9 of this book.

At a more advanced level is the recently reissued classic *Pure Mathematics* of Hardy (2008), which I used as an undergraduate. It involves more manipulation of formulas than is usually required these days, but it also insists on sound foundations, with Dedekind cuts in Chap. 2.

Still at an advanced undergraduate level, but aimed at Lebesgue integration, is Bressoud (2008). This book also includes a very attractive history of the subject, woven into the mathematical development.

In addition to the book of Bressoud just mentioned, for the history of analysis I recommend Hairer and Wanner (1996) and Jahnke (2003). Hairer and Wanner is an extremely well-illustrated book, covering mainly the sixteenth to the nineteenth century. The book edited by Jahnke is a collection of articles, of which those on the nineteenth century are particularly relevant.

10.8.4   Set Theory

Set theory is, alas, not a standard undergraduate subject, so one does not find many introductory books on it. For a very entertaining introduction see *In Search of Infinity* by Vilenkin (1995), then try the *Naive Set Theory* of Halmos (1960). Although suitable for beginners, *Naive Set Theory* is nevertheless a rigorous introduction to the ZF axioms. I might also mention Stillwell (2010), for connections between set theory and logic.

It is a huge step from this level to, say, Solovay's theorem that it is consistent with ZF to assume that all sets of real numbers are Lebesgue measurable. But there is a book that will enable you to take this step when you are ready: Jech (2003). Before doing that, a good intermediate step may be reading Cohen (1966), which introduces the forcing method that made modern set theory possible.

The history of set theory has been well served by Ferreirós (1999) and Kanamori (2003). Ferreirós is a very rich and detailed history from the beginnings to the ZF axioms; Kanamori (2003) is an advanced book on large cardinals interwoven with a history of the subject. Kanamori has also written excellent historical articles on set theory, such as Kanamori (1996).

10.8.5   Axiom of Choice

It is probably best to begin studying AC through its history in *Zermelo's Axiom of Choice* by Moore (1982). Only then can one appreciate the difficulty of even noticing AC in the early days of set theory. After that, try the comprehensive books of Jech (1973) and Herrlich (2006).

References

Abbott, S. (2001). *Understanding Analysis*. New York: Springer-Verlag.

Ackermann, W. F. (1937). Der Widerspruchsfreiheit der allgemeine Mengenlehre. *Math. Ann. 112*, 305–315.

Alexandrov, P. S. (1916). Sur la puissance des ensembles mesurables B. *Comptes Rendus Acad. Sci. Paris 162*, 323–325.

Archimedes (2004). *The Works of Archimedes. Vol. I.* Cambridge: Cambridge University Press. The two books on the sphere and the cylinder, Translated into English, together with Eutocius' commentaries, with commentary, and critical edition of the diagrams by Reviel Netz.

Artmann, B. (1999). *Euclid—the Creation of Mathematics*. New York: Springer-Verlag.

Baire, R. (1899). Sur les fonctions de variables réelles. *Annali di Mat. (3) 3*, 1–123.

Banach, S. and A. Tarski (1924). Sur la décomposition des ensembles de points en parties respectivement congruents. *Fundamenta Mathematicae 6*, 244–277.

Bernoulli, D. (1753). Réflexions et éclaircissemens sur les nouvelles vibrations des cordes exposées dans les mémoires de l'académie de 1747 & 1748. *Hist. Acad. Sci. Berlin 9*, 147–172.

Bettazzi, R. (1896). Gruppi finiti ed infiniti di enti. *Acad. Sci. Torino 31*, 446–456.

Bolzano, B. (1817). *Rein analytischer Beweis des Lehrsatzes dass zwischen je zwey Werthen, die ein entgegengesetzes Resultat gewähren, wenigstens eine reelle Wurzel der Gleichung liege.* Ostwald's Klassiker, vol. 153. Engelmann, Leipzig, 1905. English translation in Russ (2004), 251–277.

Borel, É. (1895). Sur quelques points de la théorie des fonctions. *Ann. Sci. École Norm. Sup. 12*, 9–55.

Borel, É. (1898). *Leçons sur la théorie des fonctions*. Paris: Gauthier-Villars.

Borel, É. (1905). Quelques remarques sur les principes de la théorie des ensembles. *Math. Ann. 60*.

Borel, É. (1950). *Leçons sur la théorie des fonctions* (4 ed.). Paris: Gauthier-Villars.

Bourbaki, N. (1939). *Éléments de Mathématique, Théorie des Ensembles*. Paris: Hermann. English translation, *Elements of Mathematics. Theory of Sets*, Hermann, Paris, 1968.

Bradley, R. E. and C. E. Sandifer (2009). *Cauchy's Cours d'analyse*. New York: Springer.

Bressoud, D. M. (2008). *A Radical Approach to Lebesgue's Theory of Integration*. Cambridge: Cambridge University Press.

Brouwer, L. E. J. (1911). Beweis der Invarianz der Dimensionenzahl. *Math. Ann. 70*, 161–165.

Cantor, G. (1872). Über die Ausdehnung eines Satzes aus der Theorie der trigonometrischen Reihen. *Math. Ann. 5*, 123–132.

Cantor, G. (1874). Über eine Eigenschaft des Inbegriffes aller reellen algebraischen Zahlen. *J. reine und angew. Math. 77*, 258–262. In his *Gesammelte Abhandlungen*, 145–148. English translation by W. Ewald in Ewald (1996), vol. II, 840–843.

Cantor, G. (1878). Ein Beitrag zur Mannigfaltigkeitslehre. *J. reine und angew. Math. 84*, 242–258.

J. Stillwell, *The Real Numbers: An Introduction to Set Theory and Analysis*, Undergraduate Texts in Mathematics, DOI 10.1007/978-3-319-01577-4, © Springer International Publishing Switzerland 2013

Cantor, G. (1883). Über unendliche, lineare Punktmannigfaltigkeiten, 5. *Math. Ann. 21*, 545–586.

Cantor, G. (1884). Über unendliche, lineare Punktmannigfaltigkeiten, 6. *Math. Ann. 23*, 453–488.

Cantor, G. (1891). Über eine elementare Frage der Mannigfaltigkeitslehre. *Jahresber. deutsch. Math. Verein. 1*, 75–78. English translation by W. Ewald in Ewald (1996), vol. II, 920–922.

Cauchy, A.-L. (1821). *Cours d'Analyse*. Chez Debure Frères. Annotated English translation in Bradley and Sandifer (2009).

Cohen, P. J. (1963). The independence of the continuum hypothesis. *Proceedings of the National Academy of Sciences 50*, 1143–1148.

Cohen, P. J. (1966). *Set Theory and the Continuum Hypothesis*. W. A. Benjamin, Inc., New York-Amsterdam.

Dedekind, R. (1872). *Stetigkeit und irrationale Zahlen*. Braunschweig: Vieweg und Sohn. English translation in: *Essays on the Theory of Numbers*, Dover, New York, 1963.

Dedekind, R. (1877). Sur la théorie des nombres entiers algebriques. *Bull. Soc. Math. France 11*, 278–288. English translation in Dedekind (1996).

Dedekind, R. (1888). *Was sind und was sollen die Zahlen?* Braunschweig: Vieweg und Sohn. English translation in: *Essays on the Theory of Numbers*, Dover, New York, 1963.

Dedekind, R. (1996). *Theory of Algebraic Integers*. Cambridge: Cambridge University Press. Translated from the 1877 French original and with an introduction by John Stillwell.

Descartes, R. (1637). *The Geometry of René Descartes. (With a facsimile of the first edition, 1637.)*. New York, NY: Dover Publications Inc. Translated by David Eugene Smith and Marcia L. Latham, 1954.

Dirichlet, P. G. L. (1829). Sur la convergence des séries trigonométriques qui servent à représenter une fonction arbitraire entre des limites données. *J. reine und angew. Math 4*, 157–169. In his *Werke* 1: 117–132.

Dirichlet, P. G. L. (1837). Über die Darstellung ganz willkürlicher Functionen durch Sinus- und Cosinusreihen. *Repertorium der Physik 1*, 152–174. In his *Werke* 1: 133–160.

du Bois-Reymond, P. (1875). Über asymptotische Werte, infinitäre Approximationen und infinitäre Auflösung von Gleichungen. *Math. Ann. 8*, 363–414.

Ebbinghaus, H.-D., H. Hermes, F. Hirzebruch, M. Koecher, K. Mainzer, J. Neukirch, A. Prestel, and R. Remmert (1991). *Numbers*. New York: Springer-Verlag. With an introduction by K. Lamotke, Translated from the second 1988 German edition by H. L. S. Orde, Translation edited and with a preface by J. H. Ewing,.

Euclid (1956). *The Thirteen Books of Euclid's Elements translated from the text of Heiberg. Vol. I: Introduction and Books I, II. Vol. II: Books III–IX. Vol. III: Books X–XIII and Appendix*. New York: Dover Publications Inc. Translated with introduction and commentary by Thomas L. Heath, 2nd ed.

Ewald, W. (1996). *From Kant to Hilbert: A Source Book in the Foundations of Mathematics. Vol. I, II*. New York: The Clarendon Press Oxford University Press.

Feferman, S. and A. Levy (1963). Independence results in set theory by Cohen's method II. *Notices Amer. Math. Soc. 10*, 593.

Ferreirós, J. (1999). *Labyrinth of Thought*. Basel: Birkhäuser Verlag.

Fibonacci (1202). *Liber abaci*. In *Scritti di Leonardo Pisano*, edited by Baldassarre Boncompagni, Rome 1857-1862. English translation *Fibonacci's Liber abaci*, by L.E. Sigler, Springer, New York, 2002.

Fourier, J. (1807). Mémoire sur la propagation de la chaleur dans les corps solides. *Nouveau Bulletin des sciences par la Société philomatique de Paris 1*, 112–116.

Fourier, J. (1822). *La théorie analytique de la chaleur*. Paris: Didot. English translation, *The Analytical Theory of Heat*, Dover, New York, 1955.

Fraenkel, A. (1922). Über den Begriff "definit" und die Unabhängigkeit des Auswahlaxioms. *Sitzungsber. Preuss. Akad. Wiss.*, 253–257. English translation in van Heijenoort (1967), 285–289.

Friedman, H. M. (1970/1971). Higher set theory and mathematical practice. *Ann. Math. Logic 2(3)*, 325–357.

Gale, D. and F. M. Stewart (1953). Infinite games with perfect information. In *Contributions to the Theory of Games, vol. 2*, Annals of Mathematics Studies, no. 28, pp. 245–266. Princeton, N. J.: Princeton University Press.

Gamow, G. (1947). *One, Two, Three, . . . , Infinity*. New York: Viking Press.

Gauss, C. F. (1816). Demonstratio nova altera theorematis omnem functionem algebraicum rationalem integram unius variabilis in factores reales primi vel secundi gradus resolvi posse. *Comm. Recentiores (Gottingae) 3*, 107–142. In his *Werke* 3: 31–56.

Gödel, K. (1931). Über formal unentscheidbare Sätze der Principia Mathematica und verwandter Systeme. I. *Monatshefte für Mathematik und Physik 38*, 173–198. English translation in van Heijenoort (1967), 596–616.

Gödel, K. (1938). The consistency of the axiom of choice and the generalized continuum-hypothesis. *Proceedings of the National Academy of Sciences 24*, 556–557.

Gödel, K. (1939). Consistency proof for the generalized continuum hypothesis. *Proceedings of the National Academy of Sciences 25*, 220–224.

Gouvêa, F. Q. (2011). Was Cantor surprised? *Amer. Math. Monthly 118*(3), 198–209.

Grassmann, H. (1847). *Die lineale Ausdehnungslehre*. Leipzig: Wiegand. English translation by Lloyd Kannenburg, *A New Branch of Mathematics*, Open Court, Chicago, 1995.

Grassmann, H. (1861). *Lehrbuch der Arithmetic*. Berlin: Enslin.

Hairer, E. and G. Wanner (1996). *Analysis by its History*. New York: Springer-Verlag.

Hales, T. C. (2007). The Jordan curve theorem, formally and informally. *Amer. Math. Monthly 114*(10), 882–894.

Halmos, P. R. (1960). *Naive Set Theory*. D. Van Nostrand Co., Princeton, N.J.-Toronto-London-New York.

Hamel, G. (1905). Eine Basis aller Zahlen und die unstetigen Lösungen der Funktionalgleichung $f(x + y) = f(x) + f(y)$. *Math. Ann. 60*, 459–462.

Hardy, G. H. (2008). *A Course of Pure Mathematics* (Centenary ed.). Cambridge: Cambridge University Press. Reprint of the tenth (1952) edition with a foreword by T. W. Körner.

Harnack, A. (1885). Über den Inhalt von Punktmengen. *Math. Ann. 25*, 241–250.

Hartshorne, R. (2000). *Geometry: Euclid and Beyond*. New York: Springer-Verlag.

Hausdorff, F. (1914). *Grundzüge der Mengenlehre*. Veit and Comp.

Hausdorff, F. (1916). Die Mächtigkeit der Borelschen Mengen. *Math. Ann. 77*, 430–437.

Havil, J. (2003). *Gamma*. Princeton, NJ: Princeton University Press.

Heath, T. L. (1897). *The Works of Archimedes*. Cambridge: Cambridge University Press. Reprinted by Dover, New York, 1953.

Heine, E. (1872). Die Elemente der Functionenlehre. *J. reine und angew. Math. 74*, 172–188.

Hermite, C. (1873). Sur la fonction exponentielle. *Comptes Rendus LXXVII.* 18–24, 74–49, 226–233, 285–293.. In his *Œuvres* 3, 150–181.

Herrlich, H. (2006). *Axiom of Choice*, Volume 1876 of *Lecture Notes in Mathematics*. Berlin: Springer-Verlag.

Hilbert, D. (1899). *Grundlagen der Geometrie*. Leipzig: Teubner. English translation: *Foundations of Geometry*, Open Court, Chicago, 1971.

Huntington, E. V. (1917). *The Continuum and other Types of Serial Order*. 2nd Edition. Harvard University Press.

Jahnke, H. N. (2003). *A History of Analysis*. Providence, RI: American Mathematical Society (AMS).

Jech, T. (1973). *The Axiom of Choice*. Amsterdam: North-Holland Publishing Co.

Jech, T. (2003). *Set Theory* (Third ed.). Berlin: Springer-Verlag.

Jordan, C. (1887). *Cours d'Analyse* (1 ed.). Paris: Gauthier-Villars.

Jordan, C. (1893). *Cours d'Analyse* (2 ed.). Paris: Gauthier-Villars.

Kanamori, A. (1996). The mathematical development of set theory from Cantor to Cohen. *Bull. Symbolic Logic 2*(1), 1–71.

Kanamori, A. (2003). *The Higher Infinite* (Second ed.). Berlin: Springer-Verlag.

Kechris, A. S. (1995). *Classical Descriptive Set Theory*. New York: Springer-Verlag.

Kelvin, L. (1889). *Popular Lectures and Addresses*. London: Macmillan.

Kuratowski, K. (1921). Sur la notion de l'ordre dans la théorie des ensembles. *Fundamenta Mathematicae*, 161–171.

Kuratowski, K. (1922). Une méthode d'elimination des nombres transfinis des raisonnements mathématiques. *Fundamenta Mathematicae*, 76–108.

Kuratowski, K. (1924). Sur l'état actuel de l'axiomatique de la théorie des ensembles. *Annales de la Société Polonaise de Mathématique 3*, 146–149. In Kuratowski (1988), 179.

Kuratowski, K. (1988). *Selected Papers*. Warsaw: PWN—Polish Scientific Publishers.

Landau, E. (1951). *Foundations of Analysis. The Arithmetic of Whole, Rational, Irrational and Complex Numbers*. Chelsea Publishing Company, New York, N.Y. Translated by F. Steinhardt from the German edition, *Grundlagen der Analysis*, originally published in 1930.

Lebesgue, H. (1902). Intégrale, longuer, aire. *Annali di matematica pura ed applicata 7*, 231–359.

Lebesgue, H. (1905). Sur les fonctions représentables analytiquement. *J. Math. Pures Appl. 6*, 139–216.

Levi ben Gershon (1321). *Maaser Hoshev*. German translation by Gerson Lange: *Sefer Maasei Choscheb*, Frankfurt 1909.

Liouville, J. (1851). Sur des classes très-étendues de quantités dont la valeur n'est ni algébrique, ni même rÃl'ductible à des irrationelles algébriques. *J. Math. pures appl 16*, 133–142.

Martin, D. A. (1975). Borel determinacy. *Ann. of Math. (2) 102*(2), 363–371.

Martin, D. A. and J. R. Steel (1989). A proof of projective determinacy. *Journal of the American Mathematical Society 2*(1), 71–125.

Moore, G. H. (1982). *Zermelo's Axiom of Choice*. New York: Springer-Verlag.

Mycielski, J. and S. Świerczkowski (1964). On the Lebesgue measurability and the axiom of determinateness. *Fundamenta Mathematicae 54*, 67–71.

Neeman, I. (2010). Determinacy in $L(\mathbb{R})$. In *Handbook of Set Theory. Vol. 3*, pp. 1877–1950. Dordrecht: Springer.

Neugebauer, O. and A. Sachs (1945). *Mathematical Cuneiform Texts*. New Haven, CT: Yale University Press.

Newton, I. (1665). The geometrical construction of equations. *Mathematical Papers 1*, 492–516.

Newton, I. (1671). De methodis serierum et fluxionum. *Mathematical Papers, 3*, 32–353.

Pascal, B. (1654). Traité du triangle arithmétique, avec quelques autres petits traités sur la même manière. English translation in *Great Books of the Western World*, Encyclopedia Britannica, London, 1952, 447–473.

Peano, G. (1887). *Applicazione Geometriche del Calcolo Infinitesimale*. Torino: Bocca.

Peano, G. (1889). *Arithmetices principia*. Torino: Bocca.

Peano, G. (1890). Sur une courbe, qui remplit toute une aire plane. *Math. Ann. 36*, 157–160.

Pringsheim, A. (1899). Grundlagen der allgemeinen Funktionentheorie. *Encyclopädie der mathematischen Wissenschften*, vol. II.1.1: 1–53.

Russ, S. (2004). *The Mathematical Works of Bernard Bolzano*. Oxford: Oxford University Press.

Sierpiński, W. (1927). Sur une classification des ensembles mesurables (B). *Fundamenta Mathematicae 10*, 320–327.

Sierpiński, W. (1938). Fonctions additives non complètement additives et fonctions non mesurables. *Fundamenta Mathematicae 30*, 96–99.

Smith, H. J. S. (1875). On the integration of discontinuous functions. *Proc. London Math. Soc. 6*, 140–153.

Solovay, R. M. (1970). A model of set-theory in which every set of reals is Lebesgue measurable. *Ann. of Math. (2) 92*, 1–56.

Steinhaus, H. (1965). Games, An Informal Talk. *Amer. Math. Monthly 72*(5), 457–468.

Steinitz, E. (1910). Algebraische Theorie der Körper. *J. reine und angew. Math. 137*, 167–309.

Stillwell, J. (2010). *Roads to Infinity*. Natick, MA: A K Peters Ltd.

Tarski, A. (1930). Une contribution à la théorie de la mesure. *Fundamenta Mathematicae 15*, 42–50.

Thomae, J. (1879). Ein Beispiel einer unendlich oft unstetigen Function. *Zeit. f. Math. und Physik 24*, 64.

van Heijenoort, J. (1967). *From Frege to Gödel. A Source Book in Mathematical Logic, 1879–1931.* Cambridge, Mass.: Harvard University Press.

Vilenkin, N. Y. (1995). *In Search of Infinity.* Boston, MA: Birkhäuser Boston Inc. Translated from the Russian original by Abe Shenitzer with the editorial assistance of Hardy Grant and Stefan Mykytiuk.

Vitali, G. (1905). *Sul problema della misura dei gruppi di punti di una retta.* Bologna: Gamberini e Parmeggiani.

von Koch, H. (1904). Sur une courbe continue sans tangente, obtenue par une construction géométrique élémentaire. *Archiv för Matemat., Astron. och Fys. 1.*

von Neumann, J. (1923). Zur Einführung der transfiniten Zahlen. *Acta litt. Acad. Sc. Szeged X. 1,* 199–208. English translation in van Heijenoort (1967), 347–354.

von Neumann, J. (1929). Über eine Widerspruchsfreiheitfrage in der axiomatischen Mengenlehre. *J. reine und angew. Math 160,* 227–241.

Wagon, S. (1993). *The Banach-Tarski Paradox.* Cambridge: Cambridge University Press.

Wapner, L. M. (2005). *The Pea & the Sun.* Wellesley, MA: A K Peters Ltd.

Weierstrass, K. (1872). Über continuirliche Functionen eines reellen Arguments, die für keinen Werth des letzeren einen bestimmten Differentialquotienten besitzen. *Königlich Preussichen Akademie der Wissenschaften.* In his *Werke,* II:71–74.

Weierstrass, K. (1874). *Einleitung in die Theorie der analytischen Funktionen.* Summer Semester 1874. Notes by G. Hettner. Mathematische Institut der Universität Göttingen.

Zermelo, E. (1904). Beweis dass jede Menge wohlgeordnet werden kann. *Mathematische Annalen 59,* 514–516. English translation in van Heijenoort (1967), 139–141.

Zermelo, E. (1908). Untersuchungen über die Grundlagen der Mengenlehre I. *Mathematische Annalen 65,* 261–281. English translation in van Heijenoort (1967), 200–215.

Zermelo, E. (1913). Über eine Anwendung der Mengenlehre auf die Theorie des Schachspiels. In *Proceedings of the 5th International Congress of Mathematicians,* vol. 2, Cambridge University Press, Cambridge. English translation in Zermelo (2010), 267–273.

Zermelo, E. (2010). *Ernst Zermelo: Collected Works/Gesammelte Werke.* Vol. I/Band I, Volume 21 of *Schriften der Mathematisch-Naturwissenschaftlichen Klasse der Heidelberger Akademie der Wissenschaften [Publications of the Mathematics and Natural Sciences Section of Heidelberg Academy of Sciences].* Berlin: Springer-Verlag. Edited by Heinz-Dieter Ebbinghaus, Craig G. Fraser and Akihiro Kanamori.

Zorn, M. (1935). A remark on a method in transfinite algebra. *Bull. Amer. Math. Soc. 41,* 667–670.

Index

J. Stillwell, *The Real Numbers: An Introduction to Set Theory and Analysis*,
Undergraduate Texts in Mathematics, DOI 10.1007/978-3-319-01577-4,
© Springer International Publishing Switzerland 2013

Printed in the United States
By Bookmasters